Design and Applications of Single-Site Heterogeneous Catalysts

Contributions to Green Chemistry, Clean Technology and Sustainability

Sir John Meurig Thomas

University of Cambridge, UK

 **World Scientific**

NEW JERSEY · LONDON · SINGAPORE · BEIJING · SHANGHAI · HONG KONG · TAIPEI · CHENNAI

Published by

Imperial College Press
57 Shelton Street
Covent Garden
London WC2H 9HE

Distributed by

World Scientific Publishing Co. Pte. Ltd.
5 Toh Tuck Link, Singapore 596224
USA office: 27 Warren Street, Suite 401-402, Hackensack, NJ 07601
UK office: 57 Shelton Street, Covent Garden, London WC2H 9HE

British Library Cataloguing-in-Publication Data
A catalogue record for this book is available from the British Library.

DESIGN AND APPLICATIONS OF SINGLE-SITE HETEROGENEOUS CATALYSTS
Contributions to Green Chemistry, Clean Technology and Sustainability

ISBN-13 978-1-84816-909-8
ISBN-10 1-84816-909-4
ISBN-13 978-1-84816-910-4 (pbk)
ISBN-10 1-84816-910-8 (pbk)

Typeset by Stallion Press
Email: enquiries@stallionpress.com

Printed by FuIsland Offset Printing (S) Pte Ltd Singapore

FOR JEHANE

Technical assistance with illustrations by

Professor K. D. M. Harris

(Cardiff University, Wales, UK)

FOREWORD

Homogeneous and enzymatic catalysis operate usually with well-defined molecular entities, and hence strategies for elucidation of the underlying elementary processes are quite straightforward. Quite in contrast, with heterogeneous catalysis the surface generally exposes sites with widely varying structure and composition, so that deeper understanding could only recently be reached along the "surface science" approach by using single crystal surfaces as model systems. However, a bridge between these two areas is offered by so-called single-site heterogeneous catalysts (SSHCs). These are porous inorganic solids with well-defined structures exposing inner surfaces with a high surface area in which the active centres are uniformly distributed without mutual interaction. Such systems gained considerable practical applications during the past years, and Sir John Meurig Thomas has been at the forefront of research in this area over many years. This monograph presents a concise comparison of these catalysts with enzymes and immobilised homogeneous catalysts, and then outlines in detail their structural and catalytic properties, as well as their application to clean and sustainable technologies. It will certainly become a milestone in the attempts to develop unifying concepts in catalysis.

Gerhard Ertl
August 2011

PREFACE

In December 2010, I was fortunate to be the recipient of the Gerhard Ertl Lecture Award, sponsored by the universities and Max Planck Institutes (MPIs) of Berlin as well as by the company BASF in Ludwigshafen. Conscious of the fact that, three years earlier, these universities and MPIs had set up what is arguably the largest group of academics in the world devoted to the study and application of catalysis — UniCat is the cluster of excellence which focuses on the pursuit of unifying concepts in catalysis — I chose as my topic "Unifying Concepts in Single-site Catalysis". In doing so, I knew that my audience would include many young students and post-doctoral researchers who might find it helpful if I were to trace both the emergence and the growth of this rapidly expanding field of study and application.

In order to comply with the limits imposed by a one-hour lecture, I concentrated overwhelmingly in my presentation on single-site heterogeneous catalysts (SSHCs), which has been my own field of speciality for over two decades. It was simply not possible to delve into the realm of single-site homogeneous catalysis, where the work of pioneers such as Grubbs, Schrock, Kaminsky, Britzinger, Noyori, Sharpless, Heck, Negishi, Suzuki, Stevens, Marks and others involving liquid-phase homogeneous catalysis, or the elegant gas-phase studies of Helmut Schwarz and his colleagues, stands supreme.

I attempted in my lecture to illustrate the way in which inorganic solid-state chemistry constitutes a powerful route to the design, preparation and characterization of new solid catalysts, especially those required for the era of sustainability, clean technology and green chemistry, all of which are key features of current science and the manner in which it is applied for the public good.[1]

This monograph includes several illustrations and sub-sections that could not be presented in my lecture, owing to the strictures of time. Some facets of SSHCs that are discussed rather fully here were alluded to almost subliminally during the lecture itself. The scientific principles of the various techniques used for the preparation, and especially for the *in situ* or *ex situ* characterization of new catalysts, were not covered in the lecture itself, and neither do they figure at all eminently in this monograph. We are more interested here in charting the fundamental aspects of, and the results that arise from, the design of new, potentially important and practical examples of heterogeneous catalysts, through the engineering of active sites. It has not been possible to do justice to all the active workers in this field, nor to the numerous relevant publications pertaining to it. Ever since I occupied the chair created for Michael Faraday (my scientific hero) and worked in his laboratory (and lived in his residence) at the Royal Institution (RI), London, I have felt that his practice of collecting many of his key papers and publishing them in a "united" text[2] was an admirable one. It is invidious even to suggest any comparisons with Faraday, whom Lord Rutherford described as "the greatest discoverer ever", but it is surely not ignoble to emulate him in the practice of collating and unifying one's thoughts on particularly important scientific topics. Whilst pursuing research at the Davy Faraday Research Laboratory of the RI (1986–2006), I also read with awe the publications of Faraday's friend, my fellow countryman, W. R. Grove, the inventor of the fuel cell. In his book *The Correlation of Physical Forces*,[3] Grove (who, with others, is accredited with the discovery of the first law of thermodynamics) writes: "Every one is but a poor judge where he is himself interested, and I therefore write with diffidence; but it would be affecting an indifference which I do not feel if I did not state that I believe myself to have been the first who introduced this subject as a general system…". The idea of pursuing single-site heterogeneous catalysis has occupied my mind with ever-growing intensity since I first started working at the RI.[4]

I am especially grateful for the encouragement given to me by members of my Berlin audience, particularly Professors Gerhard Ertl, Matthias Dries, Joachim Sauer and the following directors of the Fritz-Haber-Institute der Max-Planck-Gesellschaft, Professors Robert Schlögl, Hajo Freund and Gerard Meijer, to put into print an amplification of the quintessential features of my lecture. I make many other acknowledgements (see ensuing pages) to individuals who have collaborated with me in many ways. I particularly wish to single out Professor Kenneth Harris of Cardiff University for the enormous help he has provided in bringing this monograph to fruition. I am also deeply grateful to Mrs Linda Webb for her expert typing and Mr Nathan Pitt for invaluable help with the illustrations. I greatly appreciate the assistance of Professor Kenneth Harris and Ms Caroline Hancox for help with the cover. Ms Jacqueline Downs is thanked for her expert editorship. Professors Peter Edwards and Alan Windle are also thanked for their general support over many years. Last but not least, I thank my wife, Professor Jehane Ragai, for her constant help and understanding.

For convenience my text is divided into three parts. The first recalls the essence of SSHCs and why they offer a strategy for the design of new catalysts. The second describes microporous SSHCs and the third describes mesoporous ones. (In the IUPAC definitions, those nanoporous solids that have pore diameters up to $20\,\text{Å}$ ($2\,\text{nm}$) are microporous, whereas in mesoporous ones pore diameters exceed $20\,\text{Å}$.) I hope that readers will find it pedagogically and otherwise helpful that full titles are given of all the articles that appear in the references at the end of each chapter; titles are an important feature of all published work.

John Meurig Thomas, FRS, FREng.
Cambridge
June 2011

References

1. J. M. Thomas. The principles of solid state chemistry hold the key to the successful design of many heterogeneous catalysts for environmentally responsible processes. *Micropor. Mesopor. Mat.* **146**, 3 (2011).
2. Michael Faraday's *Experimental Researches in Chemistry and Physics.* Foreword to Bicentennial Edition (1991) by J. M. Thomas. Taylor and Francis, London (original publication: 1859).
3. W. R. Grove. *The Correlation of Physical Forces*, 5th ed., Longmans, Green & Co., London (1874).
4. J. M. Thomas. Uniform heterogeneous catalysts: the role of solid-state chemistry in their development and design, *Angew. Chem. Int. Ed.*, 27, 1673 (1988).

ACKNOWLEDGEMENTS

During the evolution of my work on single-site heterogeneous catalysts, the following graduate students and post-doctoral workers (many of whom are now recognized international authorities) played pivotal roles.

M. A. Alario-Franco	R. H. Jones	F. Rey
M. A. Anderson	T. Khimyak	G. Sankar
P. A. Barrett	J. Klinowski	R. Schlögl
R. G. Bell	M. Klunduk	D. S. Shepherd
S. T. Bromley	C. Lamberti	A. A. Shubin
B. Captain	S. O. Lee	B. Smit
J. Chen	D. W. Lewis	A. A. Sokol
F. Cora	P. J. Maddox	A. G. Stepanov
J. W. Couves	M. A. Makarova	O. Terasaki
M. Dugal	L. Marchese	W. Ueda
C. M. Freeman	T. Maschmeyer	S. Vasudevan
S. A. French	G. R. Millward	D. Waller
P. L. Gai	S. Natarajan	C. Williams
D. Gleeson	A. K. Nowak	D. J. Willock
J. M. Gonzalez-Calbet	T. J. O'Connell	P. A. Wright
G. N. Greaves	R. D. Oldroyd	Y. Xu
C. P. Grey	S. O. Pickett	S. Yashonath
K. D. M. Harris	R. Raja	W. Zhou
S. Hermans	S. Ramdas	
M. D. Jones	S. A. Raynor	

I am especially grateful to the following senior colleagues for their collaboration:

R. D. Adams

M. Amiridis

M. Anpo

C. R. A. Catlow

M. Che

A. K. Cheetham

S. Coluccia

A. Corma

P. P. Edwards

C. A. Fyfe

P. L Gai

L. F. Gladden

G. N. Greaves

H. Grönbeck

K. D. M. Harris

G. J. Hutchings

K. U. Ingold

B. F. G. Johnson

S. V. Ley

P. A. Midgley

T. Rayment

R. A. van Santen

K. Smith

D. E. W. Vaughan

R. Xu

A. Zecchina

(the late) K. I. Zamaraev

CONTENTS

PERMISSIONS

PNAS

Fig. 4.3   *PNAS,* **102,** 13732 (2005)

RSC Publishing

Fig. 1.3   *Phys. Chem. Chem. Phys.,* **11,** 2799 (2008)
Fig. 3.3   *Chem. Commun.,* 2833 (2006)
Fig. 3.4   *Dalton Trans.,* 5498 (2007)
Fig. 3.20   *Faraday Disc.,* **105,** 1 (1996)
Figs 4.8 and 4.10   *Microporous Framework Solids* (by
                                               P. A. Wright) (2009)
Fig. 4.9   *JCS Faraday Trans.,* **91,** 3975 (1995)
Fig. 5.2   *Chem. Commun.,* 448 (2006)
Scheme 6.1   *Chem. Commun.,* 3350 (2007)
Schemes 6.4 and 6.5   *Green Chem.,* **9,** 731 (2007)
Figs 6.11, 6.12 and 6.13   *Turning Points in Solid-State Materials
                                               and Surface Chemistry* (ed. D. M. Harris
                                               & P. P. Edwards) (2007) p. 385
Scheme 6.11, Table 6.3   *Chem. Commun.,* 1340 (1984)
   and Fig. 6.15
Fig. 7.4   *J. Mater. Chem.,* **18,** 4021 (2008)
Fig. 8.3   *Chem. Commun.,* 1126 (2003)
Figs 8.12 and 8.13   *Faraday Disc.,* **138,** 301 (2008)
Fig. 8.28   *Chem. Commun.,* 2653 (2003)

Wiley-VCH

Figs 1.1 and 3.15	*Angew. Chemie. Intl. Ed.*, **47,** 5627 (2008)
Figs 1.2, 3.18, 6.1 and 6.3	*Angew. Chemie. Intl. Ed.*, **44,** 6456 (2005)
Fig. 2.3	*Eur. J. Biochem.*, **270,** 4082 (2003)
Fig. 3.16	*Angew. Chemie. Intl. Ed.*, **37,** 2466 (1988)
Fig. 3.19(a)	*Angew. Chemie. Intl. Ed.*, **48,** 6543 (2009)
Fig. 4.2	*Angew. Chemie. Intl. Ed.*, **27,** 1673 (1988)
Fig. 4.4 and Scheme 4.1	*Handbook of Heterogeneous Catalysis,* **Vol 1** (ed. G Ertl *et al.*) (1997) p. 286
Fig. 4.7	*Angew. Chemie. Intl. Ed.*, **31,** 1472 (1992)
Fig. 4.12	*Angew. Chemie. Intl. Ed.*, **38,** 3588 (1999)
Scheme 4.2	*Green Chemistry and Catalysis* (by R. A. Sheldon & I. Arends) (2007)
Fig. 4.22	*Angew. Chemie. Intl. Ed.*, **47,** 5179 (2008)
Fig. 4.23	*Angew. Chemie. Intl. Ed.*, **50** (2011)
Fig. 4.24	*Chem. Cat. Chem.*, **2,** 1548 (2010)
Fig. 4.29	*Angew. Chemie. Intl. Ed.*, **42,** 1520 (2003)
Figs 4.31 and 4.34	*Chem. Euro. J.*, **7,** 2973 (2001)
Fig. 4.32	*Angew. Chemie. Intl. Ed.*, **39,** 2310 (2000)
Fig. 4.39	*Chem. Eur. J.*, **16,** 1368 (2010)
Fig. 4.42	*Chem. Cat. Chem.*, **3,** 95 (2011)
Fig. 5.1	*Chem. Europ. J.*, **14,** 2340 (2008)
Fig. 5.7	*Angew. Chemie.*, **37,** 1198 (1988)
Scheme 6.9	*Chem. Eur. J.*, **14,** 8098 (2008)
Figs 7.1 and 7.2	*Chem. Cat. Chem.*, **1,** 223 (2009)
Figs 7.6, 7.7 and 7.8	*Angew. Chemie. Intl. Ed.*, **46,** 1075 (2007)
Figs 8.5 and 8.6	*Angew. Chemie.*, **43,** 6745 (2004)
Fig. 8.10	*Chem. Cat. Chem.*, **2,** 402 (2010)
Fig. 8.14	*Angew. Chemie. Intl. Ed.*, **45,** 4782 (2006)
Fig. 8.25	*Chem. Cat. Chem.*, **2,** 403 (2010)

Nature (Macmillan)

Fig. 2.5	*Nature*, **309,** 589 (1984)
Fig. 3.19	*Nature*, **461,** 246 (2009)

Figs 4.6 and 4.19	*Nature,* **457,** 671 (2008)
Fig. 7.4	*Nature Chem.,* **2,** 353 (2010)

Science (AAAS)

Fig. 1.4	*Science,* **273,** 1688 (1996)
Fig. 3.9	*Science,* **317** (2007)
Fig. 5.3	*Science,* **331,** 195 (2011)
Fig. 5.7	*Science,* **328,** 602 (2010)
Fig. 8.7	*Science,* **280,** 705 (1998)
Figs 8.23 and 8.24	*Science,* **329,** 1633 (2010)

Elsevier

Fig. 2.4	*Micro. & Mesoporous Materials,* **118,** 334 (2009)
Fig. 4.13	*Micro. & Mesoporous Materials,* **146,** 1 (2011)
Scheme 4.3	*Catal. Today,* **60,** 227 (2000)
Fig. 4.30	*Appl. Catal. A,* **194,** 487 (2000)
Scheme 4.7 and Fig. 4.40	*Solid State Science,* **8,** 326 (2006)
Fig. 6.8	*J. Catal.,* **216,** 265 (2003)
Fig. 7.18	*J. Mol. Catal. A,* **139,** 253 (1999)
Figs 8.15, 8.16, 8.17 and 8.18	*Chem. Phys. Lett.,* **443,** 337 (2007)
Fig. 8.22	*J. Catal.,* **223,** 232 (2004)

American Chemical Society (ACS)

Fig. 2.2	*C&E News,* July 19 (2010)
Figs 3.7, 8.2 and 8.8 and Table 8.3	*Acc. Chem. Res.,* **36,** 20 (2003)
Fig. 3.17	*J. Am. Chem. Soc.,* **125,** 4325 (2003)
Schemes 3.1 and 3.2	*J. Am. Chem. Soc.,* **126,** 3052 (2004)
Fig. 4.1	*J. Phys. Chem.,* **B108,** 869 (2004)
Figs 4.35, 4.36, 4.37 and 4.38	*Acc. Chem Res.,* **34,** 191 (2001)

Fig. 4.41	*J. Am. Chem. Soc.*, **131**, 15887 (2009)
Fig. 6.7	*J. Phys. Chem. B*, **103**, 8809 (1999)
Fig. 6.10	*J. Phys. Chem. C*, **111**, 13 (2007)
Figs 7.9, 7.11, 7.12, 7.13, 7.16 and 7.17	*Accts. Chem. Rev.*, **41**, 708 (2008)
Table 8.2	*Ind. Eng. Chem.*, **42**, 1563 (2003)
Figs 8.26 and 8.27	*ACS Catalysis*, 1 (2011), DOI:10.1021/ CS100043

Springer

Fig. 3.11	*Topics in Catal.*, **40**, 65 (2006)
Fig. 3.21	*Topics in Catal.*, **53**, 859 (2010)
Fig. 4.20	*Topics in Catal.*, **52**, 1630 (2009)
Scheme 4.5	*Topics in Catal.*, **21**, 67 (2002)
Fig. 4.33	*Topics in Catal.*, **40**, 3 (2006)
Fig. 5.4	*Catal. Lett.*, **110**, 179 (2006)
Fig. 6.9	*Topics in Catal.*, **40**, 3 (2006)
Schemes 6.7 and 6.8	*Topics in Catal.*, **53**, 200 (2010)
Fig. 8.9	*Topics in Catal.*, **53**, 848 (2010)
Figs 8.29 and 8.30	*Topics in Catal.*, **54**, 588 (2011)

Blackwells Scientific Publishers

Fig. 3.8	*Perspectives in Catalysis* (ed. J. M. Thomas and K. I. Zamanaeu) (1992) p. 125
Figs 3.12 and 3.13	*Perspectives in Catalysis* (ed. J. M. Thomas and K. I. Zamanaeu) (1992) p. 147

Royal Society Publishing

Scheme 8.2	*Phil. Trans. R. Soc. A.*, **368**, 1473 (2010)

Taylor and Francis

Fig. 3.1   *Int. Rev. Phys. Chem.,* **8,** 207 (1989)
Fig. 5.8   *Catal. Rev.,* **46,** 369 (2004)

New Science Press Ltd in association with
Blackwell Publishing Ltd

Fig. 2.1   *Protein Structure and Function*
                             (ed. G. A. Petsko and D.Ringe) (2004)

PART I

BASICS AND BACKGROUND

CHAPTER 1

INTRODUCTION TO THE SALIENT FEATURES OF SINGLE-SITE HETEROGENEOUS CATALYSTS

Because single-site heterogeneous catalysts (SSHCs) offer unrivalled opportunities in the fields of green chemistry, clean technology and sustainable development, and also because they present more scope than other kinds of heterogeneous catalysts to design well-defined, atomically characterized, catalytically active centres, they have taken on great significance in the general area of solid inorganic catalysis. See for example some of the conversions given in Table 1.1. An outstanding advantage that SSHCs have over homogeneous analogues is that, by appropriate choice or manipulation of the framework or support to which the active centre is attached, considerable scope exists to engineer active sites and their immediate environment (rather as in protein and enzyme engineering). This enables one to exercise precise control over the selectivity, activity and durability (i.e. stability and lifetime) of the catalyst. In this way, regio-, shape and enantioselectivity can be designed. There is also the practical consideration that, unlike their homogeneous counterparts, SSHCs lend themselves to facile separation of product from unconverted reactant (and catalyst), and to their recycling as well as their reactivation (where necessary).

It is the combination of the practical and the fundamental that we focus upon in this monograph, which aims to unify the concepts exhibited by the many different kinds of SSHCs that are of relevance today.

To fix our ideas, we recall that SSHCs are those in which the active centres are spatially isolated from one another and are uniformly

3

Table 1.1   A selection of some of the facile conversions that may be effected under benign conditions with air or peroxide using designed inorganic SSHCs. (Many of these processes utilize no solvents).*

---

1. Toluene to benzyl alcohol, benzaldehyde and benzoic acid. Styrene to styrene oxide and the epoxidation of cyclo-olefins, α-pinene and (+)-limonene.
2. Aerobic, shape-selective oxidation of cyclohexane and *n*-hexane to adipic acid. Conversion of cyclohexene by H_2O_2 to adipic acid.
3. Regiospecific oxidation of linear alkanes (at the CH_3 termini) in O_2.
4. Bromine-free synthesis of terephthalic acid from *p*-xylene using air as oxidant.
5. Single-step production of ε-caprolactam from cyclohexanone; also a single-step process for producing the oxime of *c*-hexanone with NH_3.
6. Generation of hazardous reagents *in situ* for Baeyer–Villiger oxidation of ketones to lactones. *In situ* generation of hydroxylamine from NH_3.
7. One-step production of niacin (vitamin B_3) and other nitrogen-containing pharmaceutical chemicals.
8. One-step solvent-free hydrogenation of polyenes.
9. Asymmetric synthesis of pharmaceutical intermediates.
10. Regioselective and stereoselective allylic aminations.

---

*All the above conversions, as well as others, have been devised in the author's laboratory and several of them are discussed more fully in later chapters.

distributed over a surface which may be very large (e.g. 10^3 m^2 g^{-1}) and three-dimensional (3-D), a situation that is ideal for practical applications, or a small flat surface, as in an Au film with a surface monolayer of long molecules terminated by single sites, a situation conducive for fundamental studies. A good example of the latter kind is contained in the work of Hara *et al.*[1] (see Figure 1.1). It is seen that, because of the relatively long tethers to which are attached the active centres and also because each centre is approximately 7 to 10 Å away from its nearest neighbour, these active sites are spatially and energetically separated.

For both high-area solids, like zeolites or microporous aluminophosphates, i.e. AlPOs, modified mesoporous silicas and other open-structure solids on which specific active centres have been grafted,[2–8] and the low-area ones typified by Hara *et al.*'s materials, the situation exists where there is no interaction (no cross-talk) between the sites. Indeed they are deliberately engineered to display

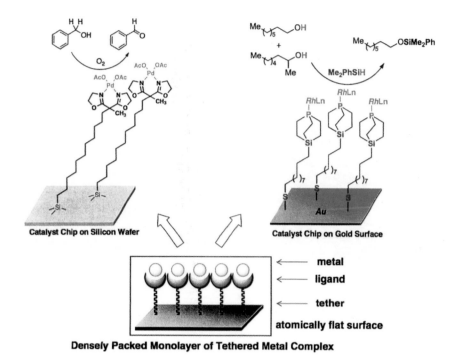

Figure 1.1   Single-site, pendant Rh-based and Pd-based catalysts attached to alkane thiolate or alkane-functionalized chains attached to low-area Au films or Si chips. (After Hara *et al.*[1])

such a property. A good example of SSHCs with well-separated centres is in H+-ZSM-5 (when the Si/Al ratio is high, see Figure 1.2).[9] Here the heat of the adsorption when a base is adsorbed on the Brønsted acid active sites remains constant with increasing coverage. In other words, all the active sites are identical.

We shall concentrate exclusively on the high-area variety of SSHCs, principally because access of reactants to the active sites (provided the pore dimensions are arranged to be large enough) is very good and a wide range of catalytic reactions can be conducted with such solid catalysts. With high-area solids, a large number of active centres may be introduced per unit mass of solid, typically from 10^{19} to 10^{21} g^{-1}, values which favour high catalytic activity if the turnover frequency at each single site is large.

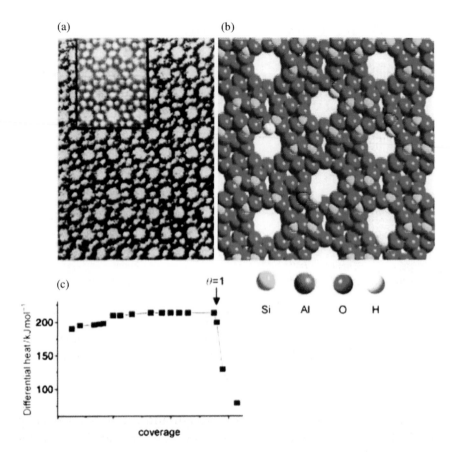

Figure 1.2 (Top left) High-resolution electron micrograph (HREM) of H-ZSM-5 acid catalyst (Si/Al ratio *ca* 25:1). Inset shows computed image. Large white spots are pores (diameter *ca* 5.5 Å) in projection. (Top right) Scalar model of the open structure showing isolated Brønsted acid sites (loosely attached hydrogen atoms, adjacent to Al[III] framework sites). (Bottom) The constancy of the heat of adsorption when pyridine is titrated against the acid sites demonstrates that these sites have the same energy and same environment. (Thomas *et al.*[9])

It is because of the current ready availability and ease of preparation of open-structure inorganic solids, represented by those shown in Figure 1.3,[6,11,12] that SSHCs have taken on such importance in the domain of pure and applied catalysis.

It is prudent to emphasize here that open-structure solids, which are the matrices to which catalytically active centres are

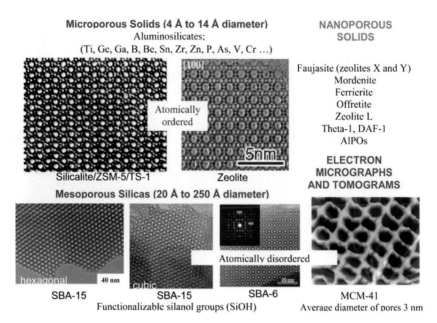

Figure 1.3 HREM photographs showing some typical microporous open-structure solid catalysts (top), where atomic order prevails, and some typical mesoporous silicas, which do not exhibit atomic order in their framework walls (bottom). (Thomas *et al.*[11])

attached, play a central role in the design and use of SSHCs. They are to be distinguished from densely packed, extended solid catalysts, such as metals, alloys and many halides, oxides and sulfides. The very concept of "active sites" (introduced in the seminal paper of H. S. Taylor[13] more than 80 years ago) is quite different for continuous solids[14–19] than it is in the open-structure solids considered here. A classic example of the kind of active sites that occur on metals is shown in Figure 1.4, which shows a monatomic step on a Ru(0001) surface. All along this step there is a continuous array of active sites, as revealed here in a scanning tunneling microscope (STM) study of what happens when NO molecules impinge upon this surface.

Some 6 min after exposure to a small quantity of gaseous NO, the STM image exhibits dark triangular spots concentrated along the step and arising from adsorbed N_{ad} species, while O_{ad} is much

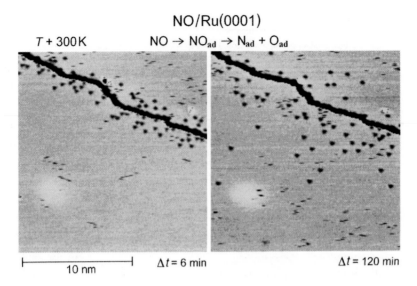

Figure 1.4   STM images from NO interacting with a Ru(0001) surface exhibiting a monatomic step, which is, effectively, a linear array of active sites (see text). (After Zambelli *et al.*[18])

more mobile and diffuses rapidly away from the step and becomes discernible as weak streaks. After 120 min or so the N_{ad} atoms have spread more across the terraces. Dissociation clearly occurs preferentially at the active sites at the step. In fact, the activation energy for dissociation of NO is a factor of 10 less at the step than at the flat surfaces.

Contrast the situation between Figure 1.4 and Figure 1.2, where the active sites created by the replacement of Si by Al in the SSHC known as the zeolite ZSM-5 are some 12 to 20 Å apart. They are spatially separated, whereas at the Ru(0001) monatomic steps they are juxtaposed, and separated by some 2 to 3 Å.

Before we review their characteristics and future potential it is instructive, first, to recall the potential of other well-known examples of single-site catalysts, namely enzymes, since much may be gleaned from what has been accomplished in the study of these biological catalysts.

References

1. K. Hara, R. Akiyama, S. Takakusagi, K. Uosaki, T. Yoshino, H. Kagi and M. Sawamura. Self-assembled monolayers of compact phosphanes with alkanethiolate pendant groups: remarkable reusability and substrate selectivity in Rh catalysts, *Angew. Chem. Int. Ed.*, **47**, 5627 (2008).
2. J. M. Thomas. Uniform heterogeneous catalysts: the role of solid-state chemistry in their development and design, *Angew. Chem. Int. Ed.*, **27**, 1673 (1988).
3. J. M. Thomas. New microcrystalline catalysts, *Phil. Trans. R. Soc. Lond. A*, **333**, 173 (1990).
4. J. M. Thomas. Design, synthesis and *in situ* characterization of new solid catalysts, *Angew. Chem. Int. Ed.*, **38**, 3588 (1999).
5. J. Čejka, H. van Bekkum, A. Corma and F. Schüth (eds). *Introduction to Zeolite Science and Practice*, Studies in Surface Science and Catalysis, Vol. 168, Elsevier, Amsterdam (2007).
6. P. A. Wright. *Microporous Framework Solids*, RSC Publishers, London (2009).
7. R. Xu, Z. Gao, J. Chen and W. Yan (eds). *From Zeolites to Porous MOF Materials*, Studies in Surface Science and Catalysis, Vol. 170, Elsevier, Amsterdam (2007).
8. J. Čejka, A. Corma and S. I. Jones (eds). *Zeolites and Catalysis: Synthesis, Reactions and Applications*, Wiley-VCH, Weinheim (2010).
9. J. M. Thomas, R. Raja and D. W. Lewis. Single-site heterogeneous catalysts, *Angew. Chem. Int. Ed.*, **44**, 6456 (2005).
10. D. J. Parrillo, C. Lee and R. J. Gorte. Heats of adsorption for NH_3 and pyridine on H-ZSM-5: evidence for identical Brønsted acid sites, *Appl. Catal. A*, **110**, 67 (1994).
11. J. M. Thomas, J. C. Hernandez-Garrido, R. Raja and R. G. Bell. Nanoporous oxidic solids: the confluence of heterogeneous and homogeneous catalysis, *Phys. Chem. Chem. Phys.*, **11**, 2799 (2008).
12. See Catalysis by Metal-Organic Frameworks: Quo Vadis? in special issue of *Phys. Chem. Chem. Phys.*, **13** (2011).
13. H. S. Taylor. Active sites in heterogeneous catalysts, *Proc. R. Soc. Lond. A*, **108**, 105 (1925).
14. K. W. Kolasinski. *Surface Science: Foundations of Catalysis and Nanoscience*, 2nd ed., John Wiley & Sons, Chichester (2007).

15. J. M. Thomas and W. J. Thomas. *Principles and Practice of Heterogeneous Catalysis*, Wiley-VCH, Weinheim (1998).
16. A. Nilsson, L. G. M. Pettersson and J. K. Nørskov. *Chemical Bonding at Surfaces and Interfaces*, Elsevier, Amsterdam (2008).
17. G. Ertl. *Reactions at Solid Surfaces*, John Wiley & Sons, Hoboken, NJ (2009).
18. T. Zambelli, J. Wintterlin, J. Frost and G. Ertl. Diffusion and atomic hopping of N atoms on Ru(0001) by scanning tunneling microscopy, *Science*, **273**, 1688 (1996).
19. J. Rostrup-Neilsen and J. K. Nørskov. Step sites in syngas catalysis, *Top. Catal.*, **40**, 45 (2006).

CHAPTER 2

LESSONS FROM THE BIOLOGICAL WORLD: THE KINSHIP BETWEEN ENZYMES AND SINGLE-SITE HETEROGENEOUS CATALYSTS

2.1 The Story of Lysozyme and Its Consequences

Very many enzymes are catalytically active even in the solid state. This being so, X-ray crystallographic studies carried out on a particular enzyme both in the absence and in the presence of a reactant (termed "substrate" by biologists) can reveal the precise mode of operation of the enzyme. It was D. C. Phillips and his associates, especially L. N. Johnson,[1,2] in the mid-to-late 1960s, who initiated such investigations when they focused on lysozyme (see Figure 2.1).

Lysozyme hydrolyses the acetal links between monomers in certain carbohydrate polymers. The substrate shown in Figure 2.1 is part of a polymer of *N*-acetyl glucosamine (NAG for short) and *N*-acetyl muramic acid (NAM). The active site of the lysozyme is a cleft that contains two key amino acids, aspartate and glutamate, labelled Asp52 and Glu35 respectively (the numbers referring to the positions of these amino acids in the chain of 129 of which lysozyme is composed). In its protonated form Glu35 acts as a weak acid and donates a proton to the carbohydrate-C-O-R- group (where R is the next sugar in the chain), breaking the C-O bond. Aspartic acid 52 in its ionized form functions as a means of stabilizing the positive charge in the transition state that builds up on the sugar (forming an oxo-carbenium ion) during catalysis. The arrows show the movement of electron pairs as bonds are made and broken.

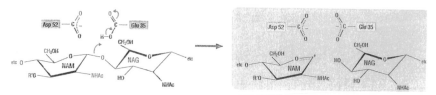

Figure 2.1   Active site of lysozyme. The enzyme lysozyme hydrolyses the acetal links between monomers in certain carbohydrate polymers. The substrate shown here is part of a polymer of *N*-acetyl glucosamine (NAG) and *N*-acetyl muramic acid (NAM). Two carboxylic acid side chains (aspartate and glutamate; purple) are found in the active site of lysozyme. In solution, these residues would be expected to have a pK_a around 4 or 5, just like acetic and lactic acids. But in the microenvironment provided by the protein, their acidities differ considerably. Aspartic acid 52 has the expected pK_a, so at pH 7 it is ionized and can fulfil its function, which is to use its negative charge to stabilize the positive charge that builds up on the sugar during catalysis. Glutamic acid 35, however, is in a hydrophobic pocket, which raises its pK_a to around 7. In its protonated form it acts as a weak acid and donates a proton to the sugar-C-O-R group (where R is the next sugar in the chain), breaking the C-O bond. In its negatively charged form it helps stabilize the positive charge on the transition state. There is recent evidence that the mechanism may also involve a covalent intermediate between aspartic acid 52 and the substrate. The red arrows show the movement of electron pairs as bonds are made and broken. (After Petsko and Ringe.[23])

    This work of Phillips led in turn to the era of protein engineering, especially after Smith had discovered the technique of site-directed mutagenesis,[3] which is a method in which specific amino acids in a protein are replaced by others. This means that the catalytic performance of an enzyme can be altered by modifying the amino acids (usually a triad of them, not just a dyad as in lysozyme) that line the pocket which constitutes the active site.

    Whilst it is not yet possible to produce modified natural (i.e. wild) enzymes that surpass the efficiency of existing ones, it is possible to build a fundamentally new enzyme. The first ever stereoselective, bimolecular (intermolecular) Diels–Alderase, which effects the combination of the two moieties depicted in Figure 2.2, was reported in 2010 by Baker and his associates in Seattle.[4] This work used the so-called Rosetta computational design methodology which makes it possible to prepare *in silico* models of the shape that is needed to

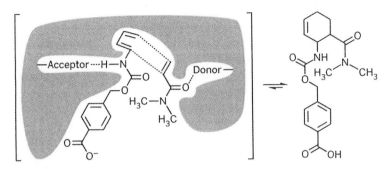

Figure 2.2 Illustration of how Baker's team (see text) envisaged an enzyme that could catalyse the intermolecular Diels–Alder reaction of the diene 4-carboxybenzyl *trans*-1,3-butadiene-1-carbonate and the dienophile *N,N*-dimethylacrylamide. (From Halford.[4])

accommodate the transition state for a particular enzymatic reaction (in this case a Diels–Alder addition). These workers then added (*in silico*) the necessary amino acids that would hold the reactants in place, an H-bond acceptor to interact with NH on the diene carbamate and an H-bond donor to interact with the dienophile carbonyl.

It is also possible by mutagenesis to enlarge the space that contains the active centres in a wild enzyme and thereby to improve greatly the catalytic performance of the resulting modified enzyme in, for example, the acylation of large asymmetrical secondary alcohols. This was done by Hult *et al.*[5] on the *Candida antarctica* lipase B, where the stereospecific pocket was modified by replacing the tryptophan 104 residue present in the wild enzyme with alanine.

It is relevant to emphasize that some important organic reactions (unlike the Diels–Alder one) cannot as yet be carried out enzymatically. One of these is the Grubbs–Schrock-type metathesis, which can, at present, be effected only by classical organometallic chemistry involving, in the case of the Grubbs catalyst, R=C links.

Darwinian evolution of enzymes is now a major research activity in biological catalysis and it is on the verge of commercial exploitation. The principles involved are shown in Table 2.1.

This approach circumvents one's profound ignorance of how the amino acid sequence encodes protein function and exploits the ability of biological systems to evolve and innovate. As Arnold,[6] one

Table 2.1 Biocatalysis by directed Darwinian evolution.

1. First select a particular wild-type enzyme (e.g. lipase from *Pseudomonas aeruginosa* or P450 from *Pseudomonas putida*).
2. The gene that encodes the enzyme is first subjected to random mutagenesis: a library of mutant genes is created in a test tube.
3. The library is then inserted into a suitable micro-organism and expressed as a library of mutant enzymes (screened for their activity and sometimes for their enantioselectivity).
4. Inferior mutants are discarded. Mutant genes of the enzyme exhibiting optimal activity (or enantioselectivity) are then subjected to further cycles of mutagenesis, expression and screening.
5. In each round, *ca* 2,000 to 5,000 mutant enzymes are created. Therefore, high-throughput screening (e.g. some 1,000 to 10,000 *ee* values to be determined per day) is required.

of the key workers in this field (Reetz is another[7]), put it, structure-guided recombination of homologous proteins generates libraries of diverse sequences, a large fraction of which retain the parent fold.

Particular examples of significant successes in this area are the conversion of glucose into isobutanol (by Arnold *et al.*), which has great promise as a biofuel; and an enantioselective cyclohexanone monooxygenase for the Baeyer–Villiger oxidation of 4-hydroxycyclohexanone to form two lactone enantiomers (by Reetz *et al.*).

Just as one knows that SSHCs may drive the next generation of inorganic catalysts towards a greener and cleaner world, so it appears that directed (Darwinian) evolution may drive the next generation of biocatalysts. As Turner recently pointed out,[8] enzymes are already increasingly used as biocatalysts for the production of chemical species that were hitherto derived using traditional (often environmentally aggressive) processes. Such products range from pharmaceutical and agrochemical building blocks to fine and bulk chemicals. Very recently, components of biofuels have been produced this way. For a biocatalyst to be effective in an industrial context, it must be subjected to improvement and optimization. In this regard, directed evolution, as outlined here, has emerged as a powerful enabling technology. To be sure, directed evolution involves

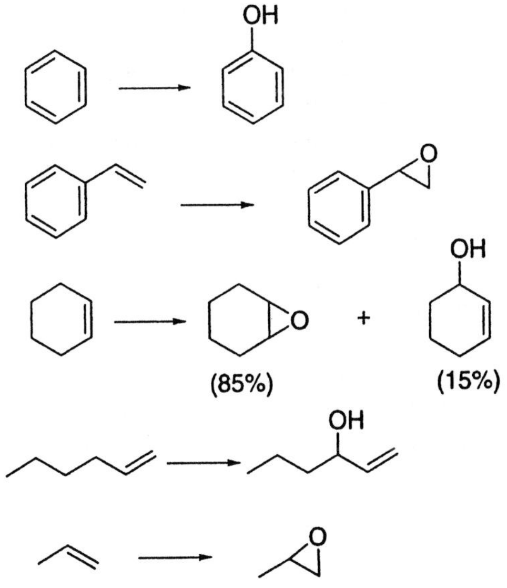

Figure 2.3 A selection of hydrocarbons that may be oxidized by cytochrome P450 enzymes. (After Fleming *et al.*[10])

repeated rounds of (i) random gene library generation, (ii) expression of genes in a suitable host and (iii) screening of libraries of variant enzymes for the property of interest. Rapid progress on all these fronts is now being made.[8,9]

A measure of the vast scope that exists for breeding or otherwise fine-tuning the structures (and hence the catalytic performance) of cytochrome P450 is reflected by the wide range of compounds that are known to be processed by such enzymes. Figure 2.3, taken from the work of Fleming *et al.*,[10] illustrates the point.

2.2 Hybrid Enzymes

A promising alternative to generating enzymes by directed evolution is to generate artificial or hybrid enzymes by inserting transition-metal-ion catalysts into the active sites of proteins, an approach pursued by Klein,[11] Ward,[12] Reetz[13] and others.[14,15] The single-site metal-ion catalyst provides the essential chemistry and first coordination sphere, whilst the protein (enzyme) provides interactions via a second coordination sphere. Asymmetric hydrogenation[2] of olefins[11,14] and enantioselective sulfoxidation are but two recent examples using this approach.

In later sections of this monograph, we shall dwell on another mode of combining SSHCs with enzymes in a procedure described as chemoenzymatic combination.[16,17]

2.3 Immobilized Enzymes

Immobilized enzymes have been studied for nearly a century, and the immobilized enzyme glucose isomerase has been used since the 1980s for the commercial production of (10^9 kg pa) fructose for the soft drinks market from corn syrup in the USA alone. In addition, the enzyme amidase, which cleaves the penicillin G or V to the penicillin nucleus, has been successfully immobilized using techniques developed by British and German multinational companies.

The advantage of encapsulating a lipase enzyme in a nanoporous system has been elegantly demonstrated by Macario *et al.*[18] as outlined in Figure 2.4. This preparation is very effective in the transesterification (with methanol) of the naturally occurring and abundant ester triolein, which is a naturally occurring glyceride of oleic acid found in fats and oils (formula $(C_{17}H_{33}COO)_3C_3H_5$), to yield very large amounts of fatty acid methyl esters (FAMEs).

2.4 The Kinship between Enzymes and SSHCs

The analogy between the mode of action of enzymes on the one hand and molecular-sieve (single-site) catalysts such as zeolites and MAPOs on the other has frequently been drawn. In a classic paper by Haag *et al.*,[19] this analogy was made explicit, with the clear evidence that Al^{III} ions (replacing Si^{IV} ones) in a microporous SiO_2 known as silicalite generated the active centres. For each framework Al^{III} ion, a loosely attached proton, as shown in Figure 2.5, is generated (i.e. a Brønsted acid site).[20] They also showed that turnover frequencies (TOFs) associated with isolated Brønsted acid sites (which give rise to the catalytic activity — see Figure 2.5) for several

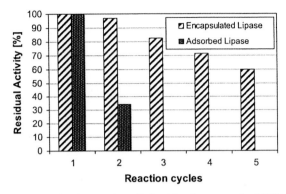

Comparison between the residual activity of the encapsulated (PAL20 sample) and adsorbed lipase after several reaction cycles (each reaction cycle: 40 °C, 18 h of reaction, triolein:methanol molar ratio 1:3 and 5 wt% of catalyst amount that means 46 mg of lipase for encapsulated enzyme and 38 mg of lipase for the adsorbed enzyme).

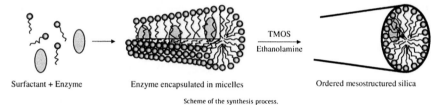

Scheme of the synthesis process.

A. Macario *et al* 2009

Figure 2.4 The stability and productivity of a lipase enzyme (in the transesterification of triolein with methanol under solvent-free conditions) are increased by encapsulation in a porous organic-inorganic system. (After Macario *et al.*[18])

hydrocarbon reactions often exceed familiar enzymatic turnover values. In later chapters we shall illustrate such analogies further (see Sections 4.3 and 4.4 in Chapter 4).

The ordinate of Figure 2.5, α, is a convenient measure of catalytic (cracking) activity. To arrive at a value of α for a particular hydrocarbon-processing catalyst, one first measures the first-order rate coefficient for the cracking of *n*-hexane at 538 °C at 100 Torr. For comparing relative activities, $\alpha = 1$ corresponds to the activity of a high-surface-area (420 $m^2 g^{-1}$) amorphous SiO_2-Al_2O_3 catalyst (10% by wt Al_2O_3), which has a first-order rate coefficient for *n*-hexane cracking of $k = 0.0675^{-1}$. This corresponds to a reaction

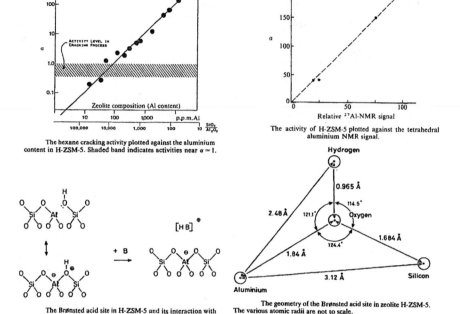

The hexane cracking activity plotted against the aluminium content in H-ZSM-5. Shaded band indicates activities near $\alpha \simeq 1$.

The activity of H-ZSM-5 plotted against the tetrahedral aluminium NMR signal.

The Brønsted acid site in H-ZSM-5 and its interaction with a base.

The geometry of the Brønsted acid site in zeolite H-ZSM-5. The various atomic radii are not to scale.

Figure 2.5 The Brønsted acid active sites in H-ZSM-5, where loosely attached protons are bound to framework oxygen atoms adjacent to the Al^{III} content (and therefore its Brønsted acid site concentration). The activity of the catalyst as a function of the tetrahedral Al content in the cracking of *n*-hexane (see text) is also shown. (After Haag *et al.*[19])

rate $(\mathrm{d}n/\mathrm{d}t = kc)$ of 1.3×10^{-7} mol s^{-1} g^{-1} of catalyst, characteristic of highly active catalyst particles in commercial petroleum cracking.

In a pair of papers[21,22] by Harris, Thomas and others in 2001, this analogy between enzymes and solid catalysts was further pursued. Thus, the solid-acid (Brønsted)-catalysed cyclodimerization of 3-hydroxy-3-methylbutan-2-one (HMB) over a synthetic (acidic) ferrierite molecular sieve was investigated. (The natural zeolite ferrierite has an empirical formula $[Mg_2Na_2(H_2O)_n][Al_6Si_{30}O_{72}]$.) HMB is a stable liquid at ambient temperatures but in acidic solution it readily undergoes reaction to generate a variety of products. However, in the acidic molecular sieve (acid ferrierite replete with Brønsted acid sites) only one product — the cyclic dimer — is observed. A plausible, proton-catalysed mechanism was proposed (see Figure 2.6). Harris, Thomas and others also investigated the solid-acid–catalysed trimerization of acetaldehyde: a highly selective and reversible transformation at ambient temperature occurred in acidic synthetic ferrierite. Acetaldehyde cyclotrimerizes (with 100% selectivity) at room temperature in this SSHC. The cyclic trimer is the only product, whereas in the liquid state again a broad distribution of products is obtained. No cyclotrimerization occurs on the Na$^+$-ion-exchanged form of ferrierite. This is a good example of the high

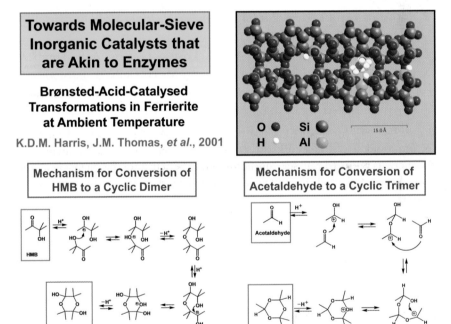

Figure 2.6 Acidic variants of synthetic versions of the zeolitic mineral ferrierite were shown by Harris, Thomas and others (see text) to be highly selective catalysts for cyclization of acetaldehyde to its trimer (right) and of 3-hydroxy-3-methylbutan-2-one (HMB) to its dimer. In acid solution a variety of oligomers is generated in both cases.

selectivity of the solid catalyst, a feature which is the hallmark of all enzymes.

The key point to note here is the obvious one that there are single-site inorganic (zeolite) catalysts that exhibit the same kind of selectivity under mild conditions of operation as enzymes.

References

1. D. C. Phillips and L. N. Johnson. Structure of some crystalline lysozyme-inhibitor complexes determined by X-ray analysis, *Nature*, **206**, 986 (1965).
2. L. N. Johnson. The early history of lysozyme, *Nat. Struct. Mol. Biol.*, **11**, 942 (1998).
3. A. J. Kirby and F. Hollfelder. *From Enzyme Models to Model Enzymes*, RSC Publishing, Cambridge (2009).
4. J. B. Siegel, A. Zanghellini, H. M. Lovick, G. Kiss, A.-R. Lambert, J.-L. St. Clair, J. L. Gallaher, D. Hilvert, M. H. Gelb, B. L. Stoddard, K. N. Houk, F. E. Michael and D. Baker. Computational design of an enzyme catalyst for a stereoselective biomolecular Diels–Alder reaction, *Science*, **329**, 309 (2010). See also B. Halford. Build your own enzyme, *C&EN*, July 19 (2010), p. 5.
5. (a) K. Hult and P. Berglund. Enzyme promiscuity: mechanism and applications, *Trends Biotechnol.*, **25**, 331 (2007).
   (b) A. Magnusson, J. C. Rotticci-Mulder, A. Santagostino and K. Hult. Creating space for large secondary alcohols by rational design of *Candida antarctica* lipase B, *Chem. Bio. Chem.*, **6**, 1051 (2005).
6. F. H. Arnold. Design by directed evolution, *Acc. Chem. Res.*, **31**, 125 (1998).
7. M. T. Reetz, A. Zonta, K. Schimossek, K. Libeton and K.-E. Jaeger. Creation of enantioselective biocatalysts for organic chemistry by *in vitro* evolution, *Angew. Chem. Int. Ed.*, **36**, 2830 (1997).
8. N. J. Turner. Directed evolution drives the next generation of biocatalysts, *Nat. Chem. Biol.*, **5**, 567 (2009).
9. J. M. Thomas and R. Raja. Designing catalysts for clean technology, green chemistry and sustainable development, *Annu. Rev. Mater. Res.*, **35**, 315 (2005).

10. B. D. Fleming, Y. Tian, S. G. Bell, L.-L. Wong, V. Urlacher and H. A. O. Hill. Redox properties of cytochrome P450 (BM3) measured by direct methods, *Eur. J. Biochem.*, **270**, 4082 (2003).

11. G. Klein. Tailoring the active site of chemzymes by using a chemogenetic optimization procedure: towards substrate-specific artificial hydrogenases based on the biotin-avidin technology, *Angew. Chem. Int. Ed.*, **44**, 7764 (2005).

12. T. R. Ward. Artificial enzymes made to order: combination of computational design and directed evolution, *Angew. Chem. Int. Ed.*, **47**, 7802 (2008).

13. M. T. Reetz, J. J.-P. Peyralans, A. Maichele, Y. Fu and M. Maywald. Directed evolution of hybrid enzymes: evolving enantioselectivity of an achiral Rh-complex anchored to a protein, *Chem. Commun.*, 4318 (2006).

14. Q. Jing, K. Okrasa and R. J. Kazlauskas. Stereoselective hydrogenation of olefins using Rh-substituted carbonic anhydrase: a new reductase, *Chem. Eur. J.*, **15**, 1370 (2009).

15. A. Pordea. Artificial metalloenzyme for enantioselective sulfoxidation based on vanadyl-loaded steptavidin, *J. Am. Chem. Soc.*, **130**, 8085 (2008).

16. P. N. R. Vennestrøm, C. H. Christensen, S. Pedersen, J.-D. Grunwaldt and J. M. Woodley. Next generation catalysis for renewables: combining enzymatic with inorganic heterogeneous catalysis for bulk chemical production, *Chem. Cat. Chem.*, **2**, 249 (2010).

17. P. N. R. Vennestrøm, E. Taarning, C. H. Christensen, S. Pedersen, J.-D. Grunwaldt and J. M. Woodley. Chemoenzymatic combinations of glucose oxidase with titanium silicalite — 1, *Chem. Cat. Chem.*, **2**, 43 (2010).

18. A. Macario, A. Corma and G. Giordano. Increasing stability and productivity of lipase enzyme by encapsulation in a porous organic-inorganic system, *Micropor. Mesopor. Mat.*, **118**, 334 (2009).

19. W. O. Haag, R. M. Lago and P. B. Weisz. The active site of acidic aluminosilicate catalysts, *Nature*, **309**, 589 (1984).

20. K. P. Schröder, J. Sauer, M. Leslie, C. R. A. Catlow and J. M. Thomas. Bridging hydroxyl groups in zeolitic catalysts: a computer simulation of their structure, vibrational properties and acidity in protonated faujasites, *Chem. Phys. Lett.*, **188**, 320 (1992).

21. S.-O. Lee, G. Sankar, S. J. Kitchin, M. Dugal, J. M. Thomas and K. D. M. Harris. Towards molecular sieve inorganic catalysts that are akin to enzymes: studies of a selective cyclodimerization over ferrierite at ambient temperature, *Catal. Lett.*, **73**, 91 (2001).

22. S.-O. Lee, S. J. Kitchin, K. D. M. Harris, G. Sankar, M. Dugal and J. M. Thomas. Acid-catalysed trimerization of acetaldehyde: a highly selective and reversible transformation at ambient temperature in a zeolitic solid, *J. Phys. Chem. B.*, **106**, 1322 (2002).

23. G. A. Petsko and D. Ringe. *Protein Structure and Function*, New Science Press, Cambridge (2004).

CHAPTER 3

DISTINCTIONS BETWEEN SINGLE-SITE HETEROGENEOUS CATALYSTS AND IMMOBILIZED HOMOGENEOUS CATALYSTS

3.1 Outline of Historical Background

From the mid-1960s and early 1970s onwards many investigators felt impelled to anchor (or graft) well-behaved homogeneous catalysts, particularly transition metal complexes (with organic ligands) and organometallic compounds, onto surfaces such as SiO_2, Al_2O_3, MgO, ZnO, TiO_2, La_2O_3, ZrO_2, Nb_2O_5, C and other continuous solids with a view to "capturing" their known single-site behaviour. Soluble metal complexes and many organometallic species are well-nigh ideal catalysts (so far as the academician is concerned) because of the wide range of chemical properties that may be incorporated into them. Thus, by invoking well-known principles of molecular design, subtle changes in electron density at, and the stereochemistry and symmetry of, the catalytically active central metal atom(s) may be readily introduced. For practical purposes, also, especially the industrial utilization of a new active catalyst, it is better to have it in a solid form, for ease of separation of products and for the purposes of recycling.

Early workers in this field were Russian scientists Yermakov and Kuznetsov[1] as well as Moiseev and Vargaftik;[2] British workers Ballard,[3] Johnson[4] and Hartley;[5] Japanese scientists especially Kuroda, Iwasawa[6] and Ichikawa;[7] US scientists Muetterties,[8] Sachtler,[9] Gates;[10] Belgian worker Jacobs;[11] van Santen[12] and others in the Netherlands; Guczi (Hungary) and Knözinger (Germany);[13] Italian investigators

Carra, Ugo,[14] Chini[15] and Bianchini;[16] Australians Masters and Maschmeyer;[17] and French workers like Che[18] and the team led by Basset[19] and later Copéret.[20] Many other workers entered this field, notably Green and Evans in the UK. Most of these studies utilized non-porous supports, and the residual hydroxyl groups of the oxides (the surface concentration of which could be controllably modified by prior thermal treatment to effect dehydroxylation) were used as anchoring points for the introduced, immobilized complexes. Japanese workers Kuroda and Iwasawa and later Ichikawa were the first to quantify the structural details of such catalysts by using X-ray absorption spectroscopy (XAFS), X-ray photoelectron spectroscopy (XPS) and thermal desorption spectroscopy (TDS). An indication of the precision with which such catalysts could be characterized is exemplified in Figures 3.1 and 3.2, taken from Iwasawa's early work.[21] Later, other spectroscopic tools, especially diffuse reflectance UV-Vis and multinuclear NMR techniques, were deployed, particularly by Bianchini, Zecchina, Coluccia and Marchese in Italy, by Che in France and by Basset and his group. This gave rise to the term "surface organometallic chemistry"[22] which initially was and still to

Yermakov, Kuroda, Iwasawa, Ballard, Basset etc

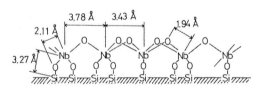

A proposed structure for Nb_2O_5 monolayer bound to the SiO_2 surface.

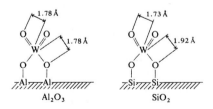

Figure 3.1   (a) Structures for a Nb_2O_5 monolayer bound to a silica surface and (b) bond distances for tungsten oxide monolayers on Al_2O_3 and SiO_2. (After Kuroda and Iwasawa.[6])

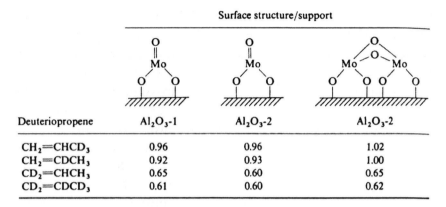

| | **Surface structure/support** | | |
Deuteriopropene	Al_2O_3-1	Al_2O_3-2	Al_2O_3-2
$CH_2{=}CHCD_3$	0.96	0.96	1.02
$CH_2{=}CDCH_3$	0.92	0.93	1.00
$CD_2{=}CHCH_3$	0.65	0.60	0.65
$CD_2{=}CDCD_3$	0.61	0.60	0.62

Figure 3.2 Kinetic isotope effects in the metathesis of propene determined by Iwasawa[21] of a molybdenum oxide monolayer on Al_2O_3 surfaces.

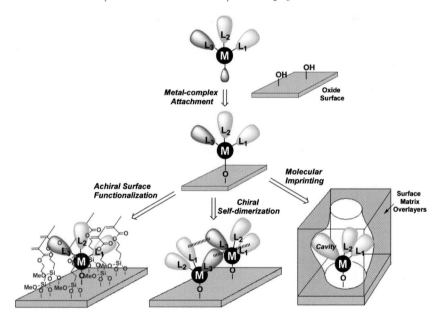

Figure 3.3 A summary of the methods employed by Tada and Iwasawa[23] for the design of supported metal complexes.

some extent is — see later section in the chapter — concerned as much with characterization as with reactivity.

Recently Iwasawa and Tada[23] have pursued this approach to an advanced degree, as may be gauged from Figure 3.3, which shows, *inter alia*, how they could attain achiral as well as chiral functionalization of metal complexes on appropriate supports. A significant advantage of SiO_2 (as we elaborate later) over, say, $\gamma\text{-}Al_2O_3$ as a support is that only one type of coordination (tetrahedral) occurs in the former, whereas for the latter (see Figure 3.4) a range of coordinations is possible,[24] which militates against the production (by anchoring) of a well-defined single-site active centre. Moreover, as Marks and co-workers have shown,[25–27] there are many different kinds of surface groups present on both partially and fully dehydroxylated aluminas. In the former, they showed that there are approximately 5.5 Lewis acidic Al^{III} centres and around 5.5 Lewis basic oxide centres per nm^2 of surface area.

There are a multiplicity of reasons why surface organometallic chemistry became so popular from the 1980s onwards:

- There appeared a powerful method of tailoring nanoparticle (supported) metal catalysts by appropriately decomposing precursor metal cluster compounds.
- The availability of synchrotron radiation (especially in the hands of early Japanese workers Kuroda, Iwasawa and Asakura) enabled precise structural studies of surface-bound organometallic molecules to be directly and quantitatively made.
- There appeared the appealing prospect of quantifying certain enigmatic aspects of bulk metal catalysts using nanoparticle

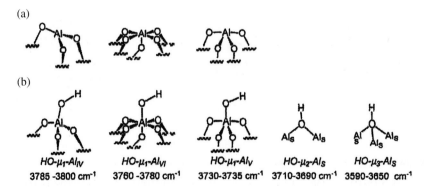

(a)

(b)

HO-μ_1-Al$_{IV}$ HO-μ_1-Al$_{VI}$ HO-μ_1-Al$_V$ HO-μ_2-Al$_S$ HO-μ_3-Al$_S$

3785 -3800 cm^{-1} 3760 -3780 cm^{-1} 3730-3735 cm^{-1} 3710-3690 cm^{-1} 3590-3650 cm^{-1}

Figure 3.4   Illustration (after Copéret[24]) of the different kinds of coordination that a Al^{III} ion may exhibit at an alumina surface.

metals for the study of adsorbed species. This was a particular feature of the pioneering work of Muetterties *et al.*[8]

- There appeared a tantalizing prospect, well illustrated in the work of Basset and co-workers, that probing the chemistry of organometallic species bound at oxide, zeolite and metal surfaces would prove illuminating in the context of understanding and designing heterogeneous catalysts.

We now proceed to elaborate on these four reasons.

3.2 Metal Cluster Compounds as Molecular Precursors for Tailored Metal Nanocatalysts

Pioneers in this field include Chini,[15] Muetterties,[8] Johnson,[4] Iwasawa and Ichikawa.[7] All these workers and others drew attention to the fact that cluster compounds, such as those shown in Figure 3.5

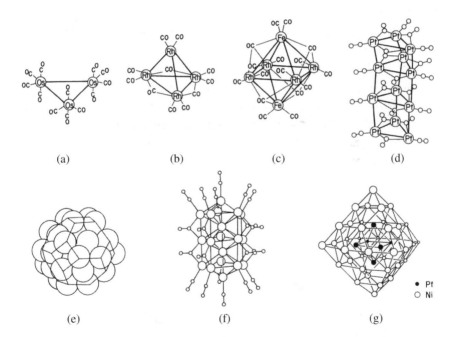

(a)   (b)   (c)   (d)

(e)   (f)   (g)

● Pt
○ Ni

Figure 3.5 Molecular clusters as precursors for tailored metal catalysts.
(a) $Os_3(CO)_{12}$; (b) $Rh_4(CO)_{12}$; (c) $Fe_2Rh_4(CO)_{16}^{2-}$; (d) $Pt_{12}(CO)_{24}^{2-}$;
(e) $[Pt_{38}(CO)_{44}H_x]^{2-}$; (f) $[Pt_{19}(CO)_{18}]^{2-}$; (g) $[Ni_{38}Pt_6(CO)_{48}H_{6-x}]^{n-}$. (After Ichikawa.[7])

possessing metal frameworks, are akin to metal nanoparticles in having neighbouring multi-metal centres.[7] All the seven structures ((a) to (g) shown here) are reminiscent of close-packed metal atoms in nanoparticles of pure metal. In particular, adjacent metal sites in polynuclear (or multinuclear, these terms are synonymous) clusters make available coordination environments that cannot be realized at a single metal atom or ionic site typical of most homogeneous catalysts. The relatively small cluster compounds — and even very small ones as shown in Figure 3.6[7] — are ideal for modelling and quantum-chemical computations. For example, it is much easier, computationally, to cope with the small metal cores (such as Fe_4C, $AuOs_3$, etc., shown in Figure 3.6) than to do a thorough computation on an extended metal structure, if one is interested in the stereochemistry and energetics of the kink site or step sites present at an Ir(100) surface.

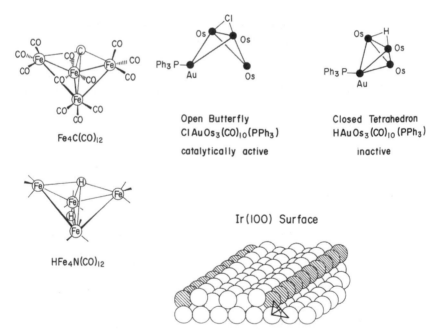

Figure 3.6 Butterfly cluster compounds useful for molecular modelling of intermediates at edge and kink sites on an Ir(100) surface. (After Ichikawa.[7])

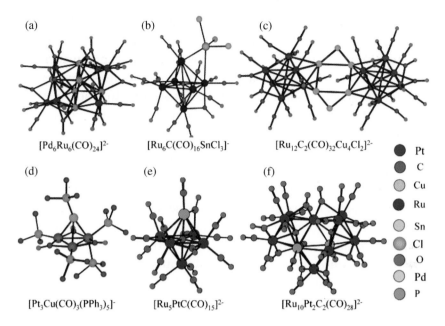

Figure 3.7 Typical parent carbonylates from which naked nanocluster (10 to 15 Å in diameter, depending upon the constituents of the bimetallic core) catalysts are generated.
(a) $[Pd_6Ru_6(CO)_{24}]^{2-}$; (b) $[Ru_6C(CO)_{16}SnCl_3]^-$; (c) $[Ru_{12}C_2(CO)_{32}Cu_4Cl_2]^{2-}$; (d) $[Pt_3Cu(CO)_3(PPh_3)_5]^-$; (e) $Ru_5PtC(CO)_{15}]^{2-}$; and (f) $[Ru_{10}Pt_2C_2(CO)_{28}]^{2-}$. (After Thomas *et al.*[28])

It is also noteworthy, as we amplify in Chapter 8, that bimetallic cluster complexes are extremely useful as precursors for the preparation of supported, minute, bimetallic nanocluster heterogeneous SSHCs (see Figure 3.7).[28] In addition to the bimetallic carbonylates shown in Figure 3.7, there are many others that have yet to be fully explored catalytically in the manner described in Chapter 8. These include $[Pd_6Fe_6(CO)_{24}]^{2-}$, $[Rh_{6-x}Fe_x(CO)_{16}]^{2-}$ with $x = 0$ to 4, and the hydro-carbonylate $[Ni_{38}Pt_6(CO)_{48}H]^{5-}$. As noted by Ichikawa,[7] fairly homogeneous mixed-phase systems such as $[Ru_6Cu(CO)_{17}]^{2-}$ and $[Rh_4Fe_2(CO)_{16}]^{2-}$ may be achieved by using mixed metal cluster complexes for metals that are immiscible in the bulk, e.g. Cu + Ru, Os + Cu and Rh + Fe.

Another important practical point is that combinations of metals in the precursor complexes having different oxophilicities may lead to anchoring of the ensemble (bimetallic nanocluster) and thereby prevent or retard sintering.[29] This fact has been demonstrated in practice in joint work by the author and Johnson *et al.*,[28,30] involving Cu_4Ru_{12} nanocluster hydrogenation catalysts.

3.3 The Essence of Surface Organometallic Chemistry (SOMC)

In a forward-looking review nearly two decades ago, Basset *et al.*,[19] following the burgeoning of organometallic chemistry in the previous three decades, outlined their definition of SOCM. *Inter alia,* they contrasted it with chemical vapour deposition (CVD) "because it deals with reactivity of the first atomic layer of a solid surface to which are attached organometallic molecules so as to obtain no more than a monolayer of a new material. By contrast CVD deals with the deposition of multilayers via the vapour phase decomposition of an organometallic compound or the vapour phase condensation of naked metal atoms or clusters".

SOMC deals with both the chemistry and the reactivity — which is not always catalytic — of surface-bound organometallic compounds such as the complexes of main group elements, the mononuclear complexes and clusters of the transition metals in different oxidation states, the complexes of lanthanides and actinides as well as others. Sometimes the object of study is to ascertain reversible adsorption (coordination) as shown in Figure 3.8;[19] sometimes the objective is to test surface (stoichiometric) reactivity[31] — not catalysis *per se* — as shown in Figure 3.9; and sometimes it is to compare catalytic performance as between a homogeneous metal complex active site with the same reaction (e.g. hydrogenation of an olefin) at a metal surface, as shown in Figure 3.10. Yet again, SOCM encompasses many novel heterogeneous catalytic reactions as summarized[32] in Figure 3.11, taken from the recent work of Taoufik *et al.* on Al_2O_3-supported tungsten hydrides as new, efficient catalysts for alkane metathesis.

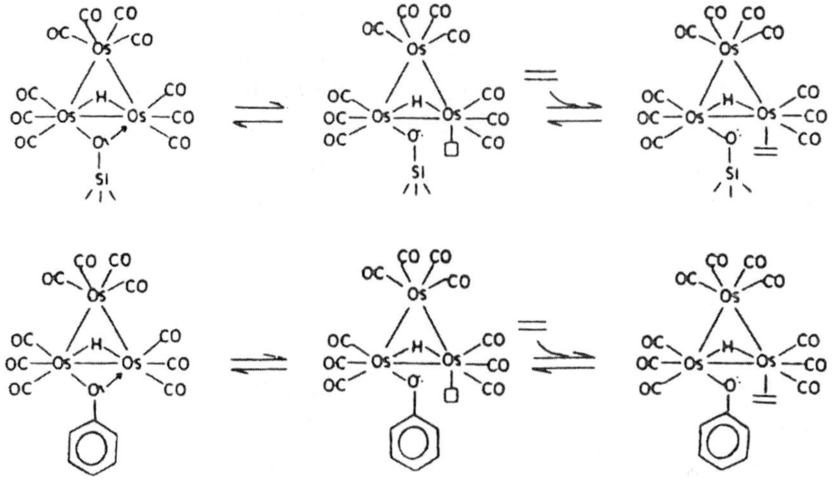

Figure 3.8 Reversible coordination of ethylene to the supported cluster (μ-H) (μ-OSi<)Os$_3$(CO)$_{10}$ and its molecular analogue (μ-H)(μ-OPh)Os$_3$(CO)$_{10}$. (After Basset *et al.*[19])

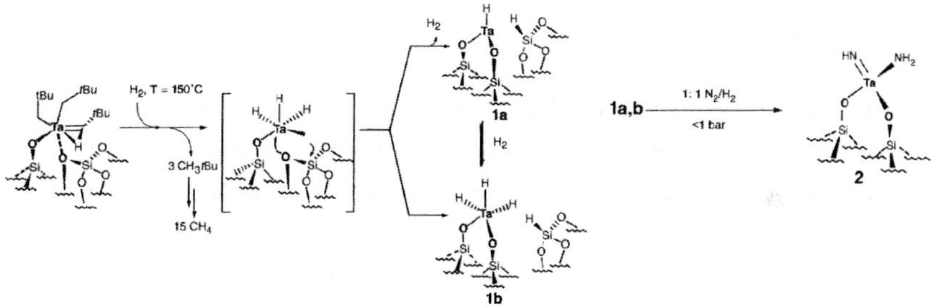

Figure 3.9 Dinitrogen dissociation on an isolated surface tantalum atom. (After Avenier *et al.*[31])

Very recently,[33] Basset and Ugo, two leading pioneers, stated that SOMC encompasses the following topics:

- Use of probe molecules on metallic surfaces as evidence of coordination and organometallic chemistry at metal surfaces.
- Chemical and structural analogy between molecular clusters and small metallic particles.

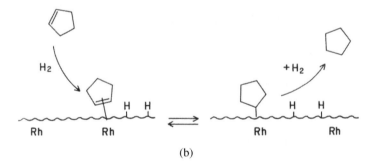

Homogeneous Metal Complex Catalysis

(a)

Heterogeneous Metal Surface Catalysis

(b)

Figure 3.10   (a) Homogeneous (metal complex) catalysis; (b) heterogeneous (metal surface) catalysis. These catalytic cycles (proposed by Ichikawa[7]) contrast the mechanistic details of olefin hydrogenation on $HRh(PPh_3)Cl$ and on an Rh surface, respectively.

- Linking organometallic surface chemistry to the elementary steps occurring on surfaces and surface stabilization of rather unstable molecular species.

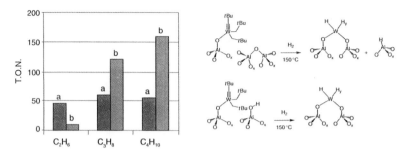

T.O.N. obtained after 120 h in the metathesis of various alkanes at 150 °C (P = 600 Torr, R = 600) catalysed by (a) [Ta]–H/SiO₂; (b) [W]–H/Al₂O₃.

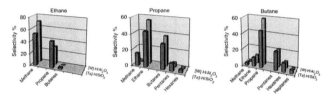

Selectivities obtained after 120 h in the metathesis of various alkanes at 150 °C (P = 600 Torr, R = 600) catalysed by [Ta]–H/SiO₂ and [W]–H/Al₂O₃.

Figure 3.11 (A) Turnover numbers obtained after 120 h in the metathesis of various alkanes at 150 °C (P= 600 Torr) catalysed by (a) [Ta]-H/SiO₂ (b) [W]-H/Al₂O₃. (B) Selectivities obtained after 120 h in the metathesis of various alkanes at 150 °C catalysed by [Ta]-H/SiO₂ and [W]-H/Al₂O₃. (After Taoufik *et al.*[32])

- Bridging surface organometallic chemistry on oxides to that on metals.
- Achieving single-metal-site heterogeneous catalysts for the design of new catalysts.

Many of these topics constitute worthy aims, but some can give rise to misleading conclusions. Take, for example, the analogies contained in Figure 3.12.[34] The structures on the right-hand side were all established by X-ray crystallography, a reliable method of determining the nature of stable (crystallizable) organometallic species. There is no doubt about the existence of π-complexes with olefins or of bridge-bonded, or threefold coordinated and fourfold (butterfly) entities, such as those illustrated more clearly in Figure 3.13.[34] But one must exercise care in drawing catalytically relevant conclusions

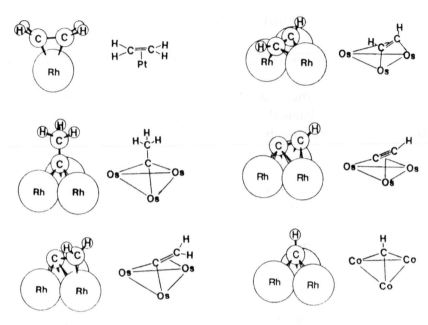

Figure 3.12 A comparison of the structures of small organic molecules on Rh surfaces and in multinuclear organometallic complexes. (After Somorjai.[34])

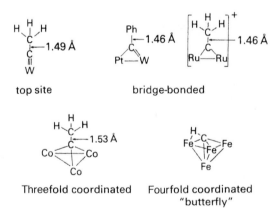

top site   bridge-bonded

Threefold coordinated   Fourfold coordinated
                          "butterfly"

Figure 3.13 Alkylidyne species detected (by LEED) on metal surface compared with those determined (by X-ray crystallography) in organometallic complexes. (After Somorjai.[34])

when similar entities (such as those shown on the left-hand side of Figure 3.12) are identified by low-energy electron diffraction (LEED). Thus Somorjai *et al.*, on detecting the ethylidyne moiety (by LEED) on a single-crystal surface of Rh, concluded that it was implicated as an intermediate in the hydrogenation of ethylene by platinum group metals. However, detailed work by Yates and Beebe,[35] using isotopic labelling and determination of turnover frequencies, showed that this entity was simply a spectator species: it does not participate in the heterogeneous catalysis.

We shall see later in this monograph that there are numerous examples of SSHCs which have little or nothing to do with SOMC. A simple illustration is that shown in Figure 3.14 where catalysis, such as the oligomerization of olefins on an H-ZSM-5 surface (which Zecchina and colleagues elucidated using *in situ* FTIR), freely ensues in the absence of a metal atom or ion.

Notwithstanding some of the deficiencies of SOMC, it is indisputable that it has proved enabling in providing satisfactory interpretations, and the language required, for many important catalytic (homogeneous) reactions.

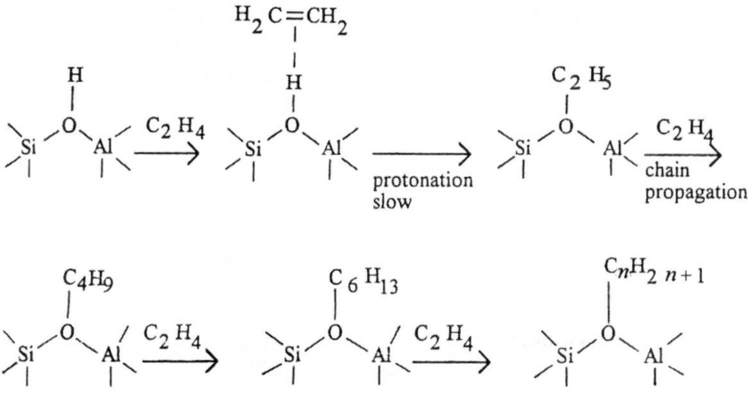

Figure 3.14 The oligomerization of ethene proceeds, catalytically, on an H-ZSM-5 surface, as measured by Zecchina *et al.* using *in situ* FTIR on the Brønsted acid active site.

3.4 Highly Active Organometallic Catalysts Based on Self-assembled Monolayers

As seen in Figure 1.1, by building an assembly of designed organo-metallic molecules on a well-defined surface (such as an Au film or Si chip), Hara *et al.*[36] developed highly efficient and reusable hetero-geneous catalysts. These workers coined the acronym SMAP (silicon-constrained monodentate alkylphosphane) for reasons that are contained in Figure 3.15. The dehydrogenative silylation of ethanol (in hexane at 25 °C)

$$EtOH + Me_2PhSiH \longrightarrow EtOSiMe_2Ph + H_2$$

proceeds rapidly and reproducibly, yielding turnover numbers as large as 235,000 in 16 h.

It is seen that, as hinted in Section 1.1, the Rh active sites are spatially separated and equivalent chemically to one another (as are

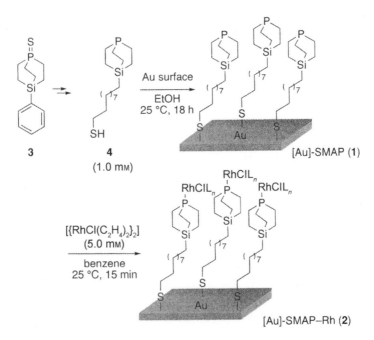

Figure 3.15 Preparation of [Au]-SMAP (1) and [Au]-SMAP-Rh (2) (see text). (After Hara *et al.*[36])

the active sites shown in Figure 1.2). Some impressive results have also been obtained by Milstein *et al.*[37] using Rh-centred Langmuir–Blodgett films tethered to glass.

3.5 Colloid-bound Organometallic Catalysts of Exceptional Activity

As is well known, colloidal metals offer great scope for surface modification — see, for example, Thomas.[38] German workers Tremel *et al.*[39] have ingeniously modified the surface of a gold colloid in such a manner as to combine the advantages of homogeneous and heterogeneous properties. By exploiting the fact that thiol-functionalized Au colloids behave like molecules — they are volatile, soluble and can be chromatographed — and possess a surface similar to that of the (111) face of bulk Au, Tremel *et al.* used a substituted long-chain thiol to tether a Ru catalyst for ring-opening metathesis polymerization (ROMP) to Au colloids. The colloid-bound Ru species heterogeneously catalyses the polymerization of norbornene in dichloromethane. At the same time, the catalyst system is readily soluble in, and recoverable from, acetone and hence behaves like a molecular species (in a homogeneously catalysed reaction). Phenomenally high turnover frequencies (typically 75,000 h^{-1}) are achieved in this way (see Figure 3.16).

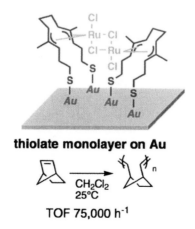

Figure 3.16 An exceptionally active polymerization catalyst (Ru-based) prepared as single sites grafted onto a thiolate monolayer on Au. (After Tremel *et al.*[39])

3.6 Analogies with Single-site Homogeneous Polymerization Catalysts

The widely used metallocene, a homogeneous single-site polymerization catalyst, involving usually a Zr ion attached to a cyclopentadiene, has also been used in an academic context, immobilized onto a sulfated alumina surface (see Figure 3.17). We do not consider such catalysts or their entirely homogeneous variants further in this account, principally because such materials, especially the exceptionally active ones associated with the work of Kaminsky, Brintzinger and Gibson (notwithstanding their great importance in polymer synthesis), do not in general contain genuine single sites. All these catalysts require a co-catalyst, such as alumoxane (rather like a co-factor in enzyme catalysis), the structure of which is not well defined and is far removed from the well-characterized single sites in the examples enumerated in Table 1.1.

When a zirconocene of general formula L_nZrR_2 (L is a cyclopentadienyl ligand and R is an alkyl group) approaches the superacidic surface of sulfated zirconia, the following cation-like structure forms:

$$L_nZrR_2 + \text{HO-support} \longrightarrow L_nZrR^+...^-\text{O-support} + RH.$$

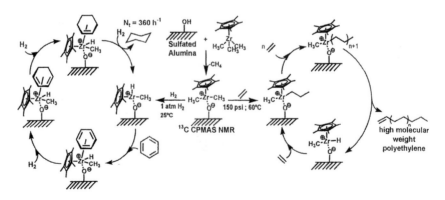

Figure 3.17 Metallocene catalysts, used widely in homogeneous solution and typified by zirconocene, are also efficient in hydrogenations and polymerizations when anchored as single sites on acidic alumina surfaces. (After Marks *et al.*[26])

This cationic-like structure is, in effect, an ion pair consisting of a cationic metallocene and a weak conjugate base of a strong Brønsted acid site. The negative charge of the conjugate base is so highly delocalized that coordination to the single-site cationic Zr centre is minimal, thereby facilitating access of the alkene to the active site.

Marks *et al.*[26] have elucidated other, related single-site organometallic electrophiles on superacidic sulfated ZrO_2, including $[Cp_2^*Th(CH_3)_2]$ and $[Cp_2^*Ti_4(CH_3)_2]$ $(Cp^* = \eta^5\text{-}C_5Me_5)$. And on sulfated Al_2O_3, they have shown that protonolysis at the strong surface Brønsted (OH) acidic site again yields a cation-like, highly reactive (in polymerization) zirconocenium electrophile $[Cp_2ZrCH_3]+$. On a fully dehydroxylated silica surface no such cation-like single-site active centre is formed.

A general method for preparing an isolated Ti-centred polymerization catalyst on porous silica was designed by McKittrick and Jones.[40] The essence of the method is depicted in Schemes 3.1 and 3.2.

Both the examples from the work of Marks *et al.*[26] and McKittrick and Jones[40] cited here are examples of SSHCs that are analogous to corresponding homogeneous single-site polymerization catalysts.

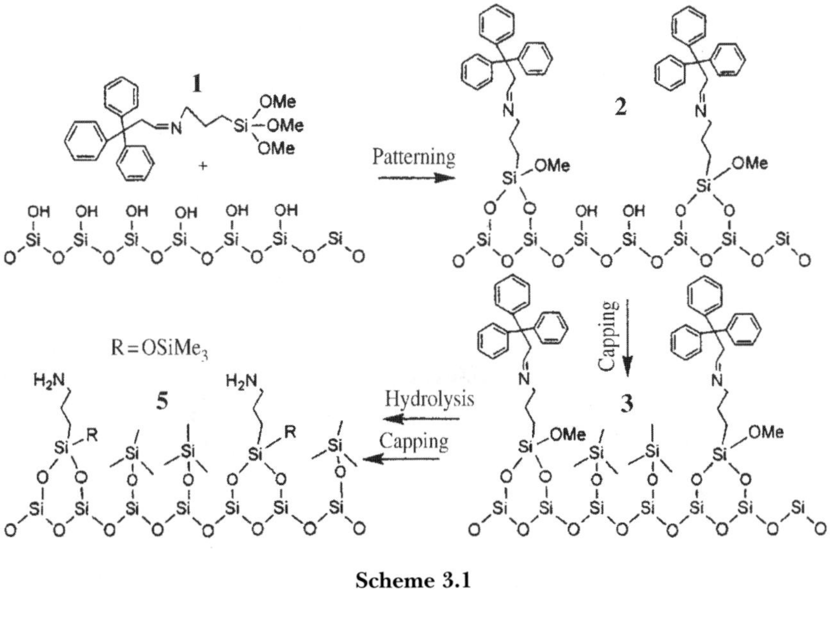

Scheme 3.1

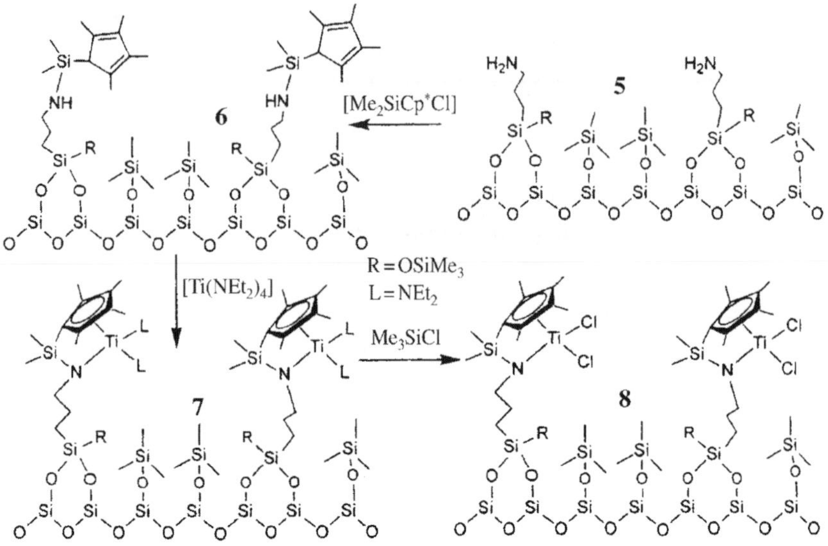

Scheme 3.2

3.7 The Taxonomy of SSHCs: A Résumé

The principal characteristic of an SSHC is that all the active sites present in it are identical in their atomic environment and hence in their energy of interaction with a reactant. The active sites are also spatially separated, thus ensuring constancy of energy of interaction with incoming species. The sites are also readily accessible to reactants since they are themselves distributed in a spatially uniform manner over the three-dimensional (3D) surface of the open-structure solids to which they are attached, either during synthesis or by subsequent modification. Figure 1.2 is an archetypal example of an SSHC. Such catalysts are in sharp contrast to those, like close-packed metals and other such solids, that possess a spectrum of active sites,[41,42] as seen both from heats of adsorption and thermal desorption measurements (see Figure 3.18), which exhibit very different thermodynamic and structural properties from those shown in Figure 1.2.

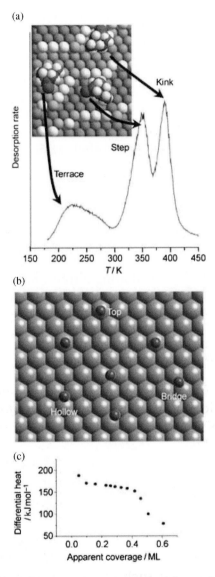

Figure 3.18 (a) and (b) A temperature-programmed desorption spectrum showing three peaks corresponding to progressively stronger bound states at terrace, step and kink sites of an adsorbate molecule (methylcyclohexanone) at a roughened Cu surface. (After Gelman.[41]) (c) The decline in the heat of adsorption with increasing coverage of a flat single-crystal surface arises because of the heterogeneity of sites and mutual repulsion of species adsorbed at neighbouring sites (ML = monolayer).[42]

There are two main categories of SSHCs which have much in common with one another. In one, a nanoporous crystalline framework of very high (3D) surface area, typically 10^3 m^2 g^{-1}, contains spatially well-separated active centres that can be readily characterized both by *in situ* and *ex situ* spectroscopic and crystallographic techniques. The top half of Figure 1.3 illustrates examples in this category: the pore diameters are less than 20 Å, and the solids are classified as microporous. In the other, again the solid matrix that accommodates the active centres is nanoporous and of very high (internal) surface area, but the matrix itself is not crystallographically ordered. However, the pores themselves (that have regular diameters in the range 20 to 500 Å) are often ordered in arrays (see lower half of Figure 1.3). These are mesoporous catalysts.

Two recently reported examples of novel kinds of SSHCs are shown in Figure 3.19.[43,44] Following a trend to increase further the accessible surface areas of microporous SSHCs,[45] Ryoo, Terasaki *et al.*[44] succeeded in preparing thin sheets (often no more than a unit cell in thickness) of H-ZSM-5 Brønsted acid catalysts. And Schlögl *et al.*[43] took advantage of the enormous surface areas of a

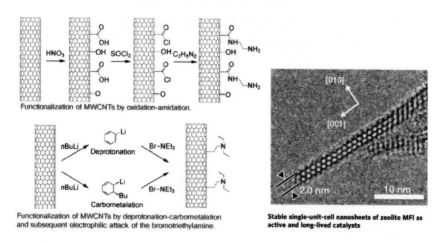

Figure 3.19 Schlögl *et al.*'s method[43] of generating SSHCs using functionalized, multi-walled carbon nanotubes (left). Ryoo *et al.*'s method[44] of producing SSHCs on high-area (single-unit-cell sheets) of ZSM-5 (designated also as MFI).

collection of carbon nanotubes, which they functionalized so as to design single-site basic heterogeneous catalysts.

The merit of using open structures, such as the microporous and mesoporous solids described, is that the wall thickness is *ca* 10 Å, which means that the so-called bulk solid is so porous that probing radiation (electromagnetic waves from IR, via visible UV to the X-ray region) can penetrate all the solids and thus interrogate species that are incarcerated (bound to the surface or rattling around) within the interior of the SSHC. This makes experimental characterization of the internal surfaces and their adsorbates conveniently possible using the traditional spectroscopic and scattering techniques of solid-state chemistry and solid-state physics[46] (see Figure 3.20).

An additional advantage in using SSHCs stems from the relative ease with which computational chemical procedures, using density functional theory (DFT) and other techniques[47] (see for example Section 4.8 in Chapter 4), may be used to elucidate many fundamental aspects of the mechanisms of catalytic change in which SSHCs are involved.

There was a recent analysis by Guidotti *et al.*[49] of the nature of SSHCs and their distinction from SOMC and other types of catalyst. Their analysis covers most of the relevant literature up until early 2010. For the sake of completeness, it should be mentioned that there are three other types of open structures that, under ideal conditions, and with appropriate pre- or post-treatment, may well turn

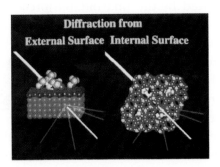

Figure 3.20 Illustration of the relative ease of probing experimentally the nature and properties of species at the internal surfaces of regular porous solids (right) compared with those at the exterior surface of a non-porous one. (After Thomas.[46])

out to be useful for the development of new kinds of SSHCs. The first are pillared clays; the second are metal-organic-framework solids (often designated as MOFs); and the third, porous organic polymers.

The author and his colleagues[50] (some of whom have pursued further such work[51,52]) explored the considerable advantages of using strongly (Brønsted) acidic clays to effect a wide range of organic conversions to which we shall return later. In principle, the act of pillaring the layers (of sheet silicates or hydrotalcites) improves the stability and lifetime of these catalysts. Moreover, the pillar itself may be appropriately functionalized so as to impart extra catalytic power to the solid.[53]

MOFs are rapidly expanding[45,54–58] kinds of hybrid solids, in which organic linker molecules are bound to a central metal ion in such a manner as to create microporous solids, some of which have enough thermal stability to withstand chemical reactions at *ca* 300 °C. In some cases, accessible active sites inside such open structures have a remarkable kinship — structurally, but not functionally (see Figure 3.21) — with certain enzymes. As pointed out by many,[57] the synthetic conditions for MOF formation are inimical to the requirements for catalytically active sites due to coordinate saturation at all metal centres (thereby eliminating metal Lewis acidity), inaccessibility of the Lewis basic sites (involved in bonding to the metal) and absence of Brønsted acid sites (candidate low pK_a moieties are deprotonated by the unavoidable presence of Lewis basic ligands). But, as Rosseinsky *et al.*,[57] Ranocchiari and van Bokhoven[58] and Lin *et al.*[59] have demonstrated, it is feasible to introduce the active site after the construction of the framework. We shall return to this topic in Chapter 7.

The third category of micro- and mesoporous solids, into which catalytically active centres may be introduced, are porous organic polymers. At the inner surface of such (disordered) polymers, functional groups, such as sulfonates or long chains terminating in carboxylic acid groups, may be introduced. It is also possible to insert free-base porphyrin sub-units. Very recently, Hupp, Nguyen *et al.*,[60] by a process of metallation with post-synthesis modification,

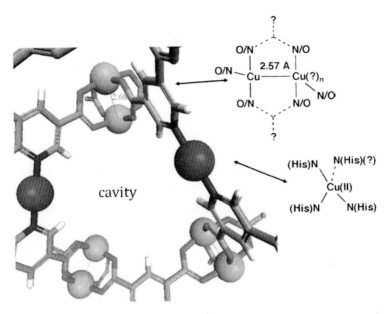

Figure 3.21 This new MOF (see text) synthesized by Norwegian workers, consisting of dimers of Cu ions, has a marked similarity to the active site in the enzyme methane monooxygenase. It is not, however, catalytically active. (After Lillerud et al.[55])

have introduced either Fe or Mn (porphyrins) that they showed to be active catalysts for both olefin epoxidation and alkane hydroxylation. It is not yet clear if the active centres in such novel catalysts are genuinely single sites in the sense described here.

References

1. Yu. I. Yermakov and B. N. Kuznetsov. *Catalysis by Supported Complexes*, Elsevier, Amsterdam (1981).
2. I. I. Moiseev and M. N. Vargaftik. Catalysis with giant palladium clusters, in *Perspectives in Catalysis*, ed. J. M. Thomas and K. I. Zamaraev, Blackwell Scientific Publications, London (1992), p. 91.
3. D. G. H. Ballard. Pi and sigma transition metal carbon compounds as catalysts for the polymerization of vinyl monomers and olefins, *Adv. Catal.*, **23**, 263 (1973).

4. B. F. G. Johnson (ed.). *Transition Metal Clusters*, Wiley, New York (1980).

5. F. R. Hartley. *Supported Metal Complexes: A New Generation of Catalysts*, D. Reidel, Dordrecht (1995).

6. H. Kuroda and Y. Iwasawa. Structural studies on catalysts and solid surfaces, *Int. Rev. Phys. Chem.*, **8**, 207 (1989).

7. M. Ichikawa. Metal cluster compounds as molecular precursors for tailored metal catalysts, *Adv. Catal.*, **38**, 283 (1992).

8. R. M. Wexler, M. C. Tsai, C. M. Friend and E. L. Muetterties. Pyridine coordination chemistry of nickel and platinum surfaces, *J. Am. Chem. Soc.*, **104**, 2034 (1982).

9. W. M. H. Sachtler and Z. Zhang. Zeolite-supported transition metal catalysts, *Adv. Catal.*, **39**, 129 (1993).

10. B. C. Gates, L. Guczi and H. Knözinger (eds). *Metal Clusters in Catalysis*, Elsevier, Amsterdam (1986).

11. A. Severeyns, D. E. De Vos and P. A. Jacobs. Towards heterogeneous and green versions of Os dihydroxylation catalysts, *Top. Catal.*, **19**, 125 (2002).

12. R. A. van Santen and M. Neurock. Concepts in theoretical heterogeneous catalytic reactivity, *Cat. Rev. Sci. Eng.*, **37**, 557 (1995) and references therein.

13. L. Guczi and H. Knözinger. Heterogeneous catalysis and C-C bond formation, *J. Mol. Catal.*, **14**, 137 (1982).

14. S. Carra and R. Ugo. Activation of covalent molecules by some noble metal complexes, *Inorg. Chim. Acta. Rev.*, **1**, 49 (1967).

15. P. Chini and G. Longini. Silica-anchored bis(trialkyl phosphine) platinum, *J. Chem. Soc. A*, 1542 (1970).

16. C. Bianchini and P. Barbaro. Recent aspects of asymmetric catalysis by immobilized chiral metal catalysts, *Top. Catal.*, **19**, 17 (2002).

17. A. F. Masters and T. Maschmeyer. Developments in silica-supported organometallic catalysts: silsesquioxanes and mesoporous MCM-41 silicates. *New Trends in Materials Chemistry*, NATO ASI Series, Vol. 498 (1997), p. 461.

18. M. Che. The molecular approach to supported catalysts synthesis: state of the art and future challenges, *J. Mol. Catal. A Chem.*, **162**, 5 (2000).

19. J. M. Basset, J. P. Candy, A. Choplin, B. Didillon, F. Quignard and A. Theolier. Surface organometallic chemistry on oxides, on zeolites

and on metals, in *Perspectives in Catalysis*, ed. J. M. Thomas and K. I. Zamaraev, Blackwell Scientific Publications, London (1992), p. 125.

20. C. Copéret, H. Adolfsson and K. B. Sharpless. A simple and efficient method for epoxidation of terminal alkenes, *Chem. Commun.*, 1565 (1997).

21. Y. Iwasawa. Chemical design surfaces for active solid catalysts, *Adv. Catal.*, **35**, 187 (1987).

22. J. M. Basset, R. Psaro, D. Roberto and R. Ugo (eds). *Modern Surface Organometallic Chemistry*, Wiley-VCH, Weinheim (2009).

23. M. Tada and Y. Iwasawa. Advanced chemical design with supported metal complexes for selective catalysis, *Chem. Commun.*, 2833 (2006).

24. C. Copéret. Design and understanding of heterogeneous alkene metathesis, *Dalton Trans.*, 5498 (2007).

25. H. Ahn and T. J. Marks. High-resolution solid-state ^{13}C NMR studies of chemisorbed organometallics, *J. Am. Chem. Soc.*, **124**, 7103 (2002).

26. C. P. Nicholas, H. S. Ahn and T. J. Marks. Synthesis, spectroscopy and catalytic properties of cationic of organozirconium adsorbates on "super acidic" sulfated alumina. "Single-site" heterogeneous catalysts with virtually 100% active sites, *J. Am. Chem. Soc.*, **125**, 4325 (2003).

27. Y. Iwasawa (ed). *Tailored Metal Catalysis*, D. Reidel, Dordrecht (1985).

28. J. M. Thomas, B. F. G. Johnson, R. Raja, G. Sankar and P. A. Midgley. High-performance nanocatalyst for single-step hydrogenations, *Acc. Chem. Res.*, **36**, 20 (2003).

29. J. M. Thomas. Design, synthesis and *in situ* characterization of new solid catalysts, *Angew. Chem. Int. Ed.*, **38**, 3588 (1999).

30. D. S. Shephard, T. Maschmeyer, B. F. G. Johnson, J. M. Thomas, G. Sankar, D. Ozkaya, W. Zhou and R. D. Oldroyd. Bimetallic nanoparticle catalysts anchored inside mesoporous silica, *Angew. Chem. Int. Ed.*, **36**, 2242 (1997).

31. P. Avenier, M. Taoufik, A. Lesage, X. Solans-Montfort, A. Baudouin, A. de Mallmann, L. Veyre, J. M. Basset, O. Eisenstein, L. Emsley and E. A. Quadrelli. Dinitrogen dissociation on an isolated surface tantalum atom, *Science*, **317** (2007).

32. M. Taoufik, E. Le Roux, J. Thivolle-Cazat, C. Copéret, J. M. Basset, B. Maunders and G. J. Sunley. Alumina-supported tungsten hydrides: new efficient catalysts for alkane metathesis, *Top. Catal.*, **40**, 65 (2006).

33. J. M. Basset and R. Ugo. On the origins and development of "surface organometallic chemistry", in *Modern Surface Organometallic Chemistry*, ed. J. M. Basset, R. Psaro, D. Roberto and R. Ugo, Wiley-VCH, Weinheim (2009).

34. G. A. Somorjai. Correlations and differences between homogeneous and heterogeneous catalysis: a surface science view, in *Perspectives in Catalysis*, ed. J. M. Thomas and K. I. Zamaraev, Blackwell Scientific Publications, London (1992), p. 147.

35. J. T. Yates and T. P. Beebe. An *in situ* infra red spectroscopic investigation of the role of ethylidyne in the ethylene hydrogenation reaction on Pd/Al_2O_3, *J. Am. Chem. Soc.*, **108**, 663 (1986).

36. K. Hara, R. Akiyama, S. Takakusagi, K. Uosaki, T. Yoshino, H. Kagi and M. Sawamura. Self-assembled monolayers of compact phosphanes with alkanethiolate pendant groups: remarkable reusability and substrate selectivity in Rh catalysis, *Angew. Chem. Int. Ed.*, **47**, 5627 (2008).

37. K. Töllner, R. Popovitz-Biro, M. Lahav and D. Milstein. Impact of molecular order in Langmuir-Blodgett films on catalysis, *Science*, **278**, 2100 (1997).

38. J. M. Thomas. Colloidal metals: past, present, future, *Pure Appl. Chem.*, **60**, 1517 (1988).

39. M. Bartz, J. Küther, R. Seshardi and W. Tremel. Colloid-bound catalysts for ring-opening metathesis polymerization: a combination of homogeneous and heterogeneous properties, *Angew. Chem. Int. Ed.*, **37**, 2466 (1988).

40. M. W. McKittrick, and C. W. Jones. Towards single-site, immobilized molecular catalysts: site-isolated Ti ethylene polymerization catalysts supported on porous silica, *J. Am. Chem. Soc.*, **126**, 3052 (2004).

41. J. D. Horvath and A. J. Gelman. Naturally chiral surfaces, *Top. Catal.*, **25**, 9 (2003).

42. J. M. Thomas, R. Raja and D. W. Lewis. Single-site heterogeneous catalysts, *Angew. Chem. Int. Ed.*, **44**, 6456 (2005).

43. J.-P. Tessonnier, A. Villa, O. Majoulet, D. S. Su and R. F. Schlögl. Defect-mediated functionalization of carbon nanotubes as a route to design single-site basic heterogeneous catalysts for biomass conversion, *Angew. Chem. Int. Ed.*, **48**, 6543 (2009).

44. M. Choi, K. Na, J. Kim, Y. Sakomoto, O. Terasaki and R. Ryoo. Designed single-site acid heterogeneous catalysts, *Nature*, **461**, 246 (2009).

45. R. E. Morris. Some difficult challenges for the synthesis of nanoporous materials, *Top. Catal.*, **53**, 1291 (2010).

46. J. M. Thomas. Catalysis and surface science at high resolution, *Faraday Disc.*, **105**, 1 (1996).

47. J. M. Thomas, C. R. A. Catlow and G. Sankar. Determining the structure of active sites, transition states and intermediates in hetero-geneously catalysed reactions (Focus article), *Chem. Commun.*, 2921 (2002).

48. C. R. A. Catlow, S. A. French, A. A. Sokol and J. M. Thomas. Computational approaches to the determination of active site structures and reaction mechanisms in heterogeneous catalysts, *Phil. Trans. R. Soc. Lond.*, **363**, 913 (2005).

49. V. Dal Santo, F. Liguori, C. Pirovano and M. Guidotti. Design and use of nanostructured single-site heterogeneous catalysts for the selective transformation of fine chemicals, *Molecules*, **15**, 3829 (2010).

50. P. A. Diddams, J. M. Thomas, W. Jones, J. A. Ballantine and J. H. Purnell. Synthesis, characterization and catalytic activity of beidellite-montmoril-lonite layered silicates and their pillared analogues, *J. Chem. Soc. Chem. Commun.*, 1340 (1984).

51. K. Chibwe and W. Jones. The synthesis of polyoxometalate-pillared layered double hydroxides as calcined precursor catalysts, *Chem. Mater.*, **1**, 489 (1989).

52. R. Mokaya and W. Jones. Pillared acid-activated clay catalysts, *Chem. Commun.*, 929 (1994).

53. J. M. Thomas and W. J. Thomas. *Principles and Practice of Heterogeneous Catalysts*, Wiley-VCH, Weinheim (1997), p. 624.

54. See P. A. Wright. *Microporous Framework Solids*, RSC Publishing, London (2008), and Refs 55 and 56 for recent reviews.

55. K. P. Lillerud, U. Olsbye and M. Tilset. Designing heterogeneous cata-lysts by incorporating enzyme-like functionalities into MOFs, *Top. Catal.*, **53**, 859 (2010).

56. G. Ferey. Metal-organic frameworks: the young child of the porous solids family, *Stud. Surf. Sci. Catal.*, **170**, 66 (2007).

57. M. J. Ingleson, J. Perez Barrio, J. Bacsa, C. Dickinson, H. Park and M. J. Rosseinsky. Generation of a solid Brønsted acid site in a chiral framework, *Chem. Commun.*, 1287 (2008).

58. M. Ranocchiari and J. A. van Bokhoven. Catalysis by metal-organic frameworks: fundamentals and opportunities, *Phys. Chem. Chem. Phys.*, **13**, 6388 (2011).

59. C. D. Wei, A. Hu, L. Zhang and W. Lin. A homochiral porous metal-organic framework for highly enantioselective heterogeneous asymmetric catalysis, *J. Am. Chem. Soc.*, **127**, 8940 (2005).

60. A. M. Shultz, O. K. Farha, J. T. Hupp and S. T. Nguyen. Synthesis of catalytically active porous organic polymers from metalloporphyrin building blocks, *Chem. Sci.*, **2**, 686 (2010).

PART II

MICROPOROUS OPEN STRUCTURES

CHAPTER 4

MICROPOROUS OPEN STRUCTURES FOR THE DESIGN OF NEW SINGLE-SITE HETEROGENEOUS CATALYSTS

4.1 Introduction

By far the most important crystalline microporous solids available for the creation of new SSHCs are zeolites or zeotypes such as aluminophosphates (AlPOs), metal aluminophosphates (MAPOs) or silico-aluminophosphates (SAPOs). At the time of writing, the most recent issue of the *Atlas of Zeolite Framework Types* (2010)[1] describes the 194 distinct microporous structures now known, all formed by corner-sharing tetrahedra TO_4 (where T may stand for Si, Al, Ge, Ga, B, Sn, P, etc.). There are some 40 naturally occurring zeolites — and very many synthetic ones, like zeolite Beta, zeolite L, zeolite Theta-1 — the general formula of which is $\frac{M_x^{n+}}{n}(Al_xSi_{1-x})O_2 \cdot mH_2O$, where x is the fraction of Si^{IV} tetrahedral sites substituted by Al^{III} ions, thereby generating x/n extra-framework ions of valence n, and m is the number of occluded water molecules. Of late (see Chapters 3 and 7) nanoporous solids known as metal-organic frameworks (MOFs) as well as covalent organic frameworks (COFs) and nanoporous polymers have come to the fore.[2] However, as yet, they play a relatively insignificant part — but see Section 7.3 where chiral MOF catalysts are discussed — in the general scheme of SSH microporous catalysis.

We saw earlier — see Figures 1.2, 1.3 and 2.5 — some typical examples of the nature of microporous zeolites, both natural and synthetic. Later in this and subsequent chapters we shall show many other catalytically significant zeotypic structures, of which, in principle (from a theoretical standpoint), many tens of thousands more[3,4]

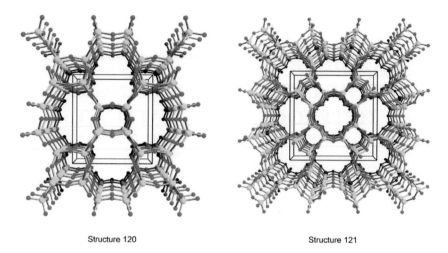

Structure 120   Structure 121

Figure 4.1 View of two (structures 120 and 121) chemically feasible but at present hypothetical crystalline networks. Structure 120 contains a one-dimensional channel system of eight-membered rings (4.8 × 4.4 Å) along [001], whereas structure 121 contains two kinds of one-dimensional channels (3.9 × 3.9 and 3.1 × 3.1 Å) of eight-membered rings along [001]. (After A. Simpeter *et al.*, *J. Phys. Chem. B,* **108,** 869 (2004).)

may exist: two of these (at present hypothetical) structures are shown in Figure 4.1.

Sometimes it is convenient to picture zeolites in terms of polyhedral sub-units[1] as in Figure 4.2,[5] where we depict zeolite X that has the same structural architecture as the zeolitic mineral known as faujasite, but with an Si/Al ratio that falls between 1.0 and 1.5. In zeolite Y, which also has the faujasite structure, the Si/Al ratio exceeds *ca* 1.6 and reaches 3 to 4, but may, by a process known as dealumination (i.e. by steaming or treatment with $SiCl_4$, and is readily monitored by solid-state ^{29}Si NMR), reach a value of infinity. Such a purely siliceous crystalline microporous solid is known as faujasitic silica. Faujasite itself (designated by the three-letter acronym FAU in the IZA convention) has a chemical formula $[(Ca,Mg,Na)_{29}]$ $[Al_{58}Si_{134}O_{384}]240H_2O$. Some or all of these extra-framework ions (Ca,Mg,Na) may be readily exchanged and replaced by other ions, e.g. La^{3+}, NH_4^+ or H_3O^+. It is very easy to synthesize faujasite zeolites

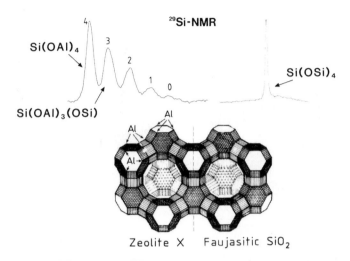

^{29}Si-NMR

Si(OAl)$_4$

Si(OAl)$_3$(OSi)

Si(OSi)$_4$

Zeolite X   Faujasitic SiO$_2$

Figure 4.2 Crystalline faujasitic silica (right) may be prepared from zeolite X or Y (left) by treatment with SiCl$_4$ (to effect dealumination). The ^{29}Si MASNMR spectrum of the product shows one peak corresponding to one type of crystalline site for the silicon in the monophasic, nanoporous silica.[5]

(X and Y) which have micropores of 7.4 Å diameter running in the <111> crystallographic directions through the cubic solid (unit cell dimension 24.3 Å).[6]

To fix our ideas further on the nature of microporous zeolytic catalysts, we illustrate in Figure 4.3 various ways of showing the active site and the structural environment of the two distinct kinds of catalyst — a Brønsted acid one and a redox one — in the siliceous structure known as silicalite-1. Silicalite has the same structure as the Brønsted acid catalyst ZSM-5 (see Figure 4.3), the latter being an aluminosilicate with its active site illustrated in the bottom right-hand corner of Figure 4.3. The presence of an Al^{III} ion in place of an Si^{IV} ion in the framework tetrahedral (T) sites generates a proton that is loosely attached to a framework oxygen. When, however, an Si^{IV} ion is replaced by Ti^{IV}, the redox catalyst TS-1 is formed (top right-hand corner of Figure 4.3). In the IZA nomenclature, silicalite-1 and TS-1 have a framework structure designated MFI. (This comes from the fact that ZSM-5 stands for zeolite Socony Mobil number five.) A closely related microporous framework known as MEL, which is

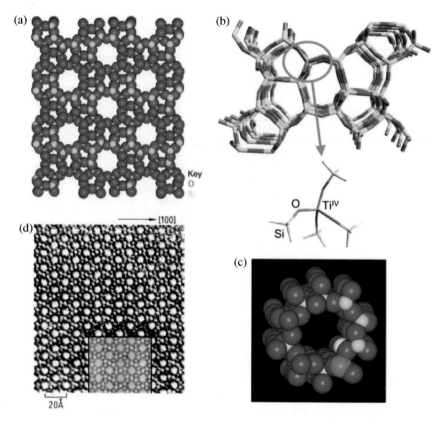

Figure 4.3 Structural drawing and projected image of redox and acid active centres in silicalite. (a) Silicalite, SiO_2, is a synthetic polymorph of silica possessing channels of pore diameter 5.5 Å. (b) TS-1 (titanosilicalite number 1) is silicalite containing a few Ti^{IV} ions as Lewis acid centres in place of some of the Si^{IV} ions that occupy the tetrahedral sites. (c) Brønsted acid centres ≡ Al-O(H)-Si ≡ are created in silicalite when Al^{III} ions (shown in green) replace some of the Si^{IV} ions (yellow) and a loosely attached proton (white) is bound to a neighbouring framework oxygen atom. (d) High-resolution micrograph of the silicalite structure imaged along the [010] direction. Inset shows computed image.

derived from ZSM-11, Mobil eleven, is also an important SSHC as we shall see later. The relationship between ZSM-5 (MFI) and ZSM-11 (MEL) is shown in Figure 4.4, where we see that a central channel, circumscribed by ten T sites, runs through each of these structures in the <010> directions.[7] It is convenient to picture each of these two

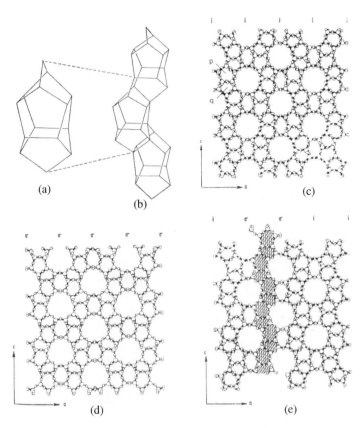

Figure 4.4 (a) Segment of the structure of ZSM-5 and ZSM-11 showing connected five-membered rings composed of linked tetrahedra (SiO$_4$ and AlO$_4$). Each connecting line represents an oxygen bridge (b), the chains from which the ZSM-5 and ZSM-11 structures are built are themselves made up by linking the units shown in (a). (c) In ZSM-5 chains are linked such that (100) slabs are related by inversion (i). (d) In ZSM-11 chains are linked such that (100) slabs are mirror images (σ) of one another. (e) Representation of intergrowths of ZSM-5 and ZSM-11. p and q refer, respectively, to the larger and smaller five-membered rings.[7]

related structures to be made up of connected strips of so-called pentasil units (see Figure 4.4(a) and (b)). In ZSM-5 (silicalite-1) the strips have been joined with inversion symmetry (i) — see Figure 4.4(c). In ZSM-11 contiguous strips are in mirror relation (σ) — see Figure 4.4(d). In typical specimens of ZSM-5 or ZSM-11 there may be sub-unit-cell intergrowths of one in the other, as seen in Figure 4.5.

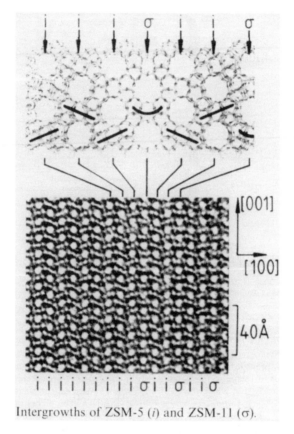

Intergrowths of ZSM-5 (*i*) and ZSM-11 (σ).

Figure 4.5 High-resolution electron micrograph (compare with Figure 4.3(d)). Typical example of a ZSM-5 specimen containing an intergrowth of ZSM-11.

Microporous aluminosilicate (and other zeotypic) catalysts possess, in effect, three-dimensional (3D) surfaces, replete with channels and cages. The channels may or may not intersect, and some (like mordenite) have side pockets. Lining the pores and distributed in more or less spatially uniform fashion throughout their bulk are the active sites which are bridging hydroxyl groups, as shown in Scheme 4.1, where Structure I represents a Brønsted acid site. Interaction with water vapour (for example) is thought to result in the physisorbed complex II, although a more plausible alternative is the H_3O^+ complex, shown as Structure III. These are the classic

Brønsted acid centres, the intrinsic strength of which is a function of both the particular local environment of the aluminosilicate structure in which they occur and also the Si:Al ratio.[8]

4.2 The Salient Characteristics of Microporous SSHCs

By adjusting the Si:Al ratio in a zeolite, and exercising the appropriate care in synthesis, it is relatively easy to ensure that the bridging hydroxyls (Structure I in Scheme 4.1), which are the source of the "free" protons that are an essential feature of Brønsted acid catalytic activity, are sufficiently far apart from one another. This is the essence of their single-site character. Such sites may line the channels and/or the cages of the zeolite, or be at their intersections. (Experimental methods, such as refined IR spectroscopy using alkanes as probes, are used[9,10] to probe the location of such sites.)

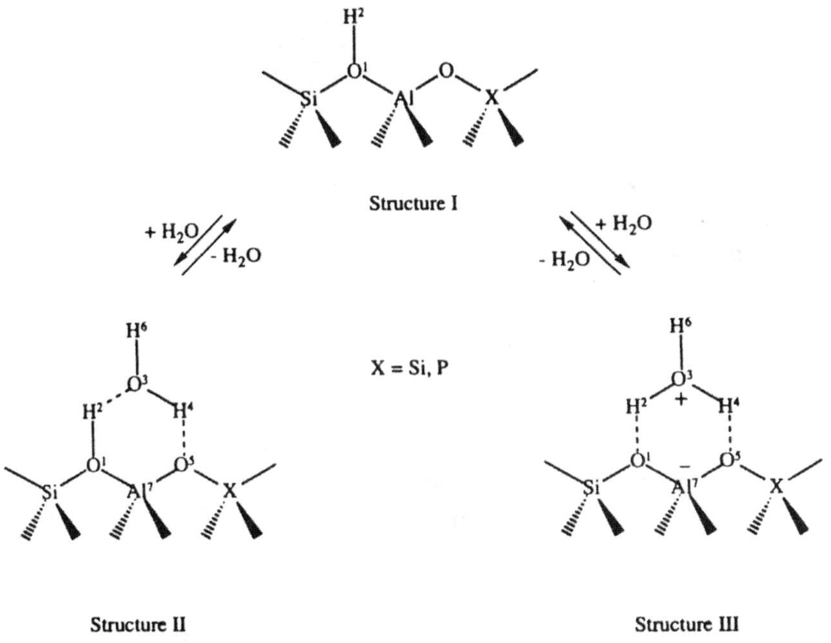

Scheme 4.1 Representation of the classic Brønsted acid centres in zeotype catalysts.

The hydrophobicity or hydrophilicity may also be controlled — see Figure 4.2, where the hydrophilic zeolite X is converted to the hydrophobic faujasitic silica on thorough dealumination. Because the host zeolite (or zeotype in the case of MAPOs and SAPOs, as shown later) is crystalline, a catalytically active site at the internal surface of such a solid has a well-defined atomic structure, which is retrievable by a number of powerful experimental methods, involving both spectroscopy and/or diffraction.

Because all these zeolites have well-defined pore diameters, they serve in a general sense as molecular sieves. The notions of shape selectivity and regioselectivity are an intrinsic feature of their properties — and shape-selective catalysis, which may manifest itself in three self-explanatory ways: reactant shape selectivity, product shape selectivity and transition-state selectivity, examples of which will be seen later (see Figure 4.16).

Zeotypic catalysts are often described as being made up of eight-membered rings or ten-membered rings, etc., abbreviated as 8MR, 10MR, 12MR, 14MR, 16MR, 18MR and 20MR, some of which (notably the main example of a 20MR zeotype, known as JDF 20) are of low thermal stability. In the main, however, catalysts such as H-ZSM-5, zeolite Beta and AlPO-5 have high (internal) surface areas, typically up to $1,000 \text{ m}^2 \text{g}^{-1}$, and are often thermally stable to just below 1,000 °C.

Finally, by appropriate introduction of heteroatoms into their frameworks a very wide range of catalytic characteristics may arise, which include acid, base and redox catalysis, and it is frequently feasible to engineer bifunctional catalysts combining typically acidic and redox properties. Figures 4.6 and 4.7 show a small selection of the more important family of AlPOs, MAPOs and SAPOs, from which, in particular, it is seen that Brønsted acidity is injected into an AlPO not only by introducing in place of Al^{III} tetrahedrally coordinated framework ions M^{II} ($= Co^{II}$, Mg^{II}, Fe^{II}, etc.), but also by introducing Si^{IV} framework ions in place of P^{IV} tetrahedrally coordinated ones. It is to be noted that just as when Ti^{IV} ions replace Si^{IV} in silicate, so also when Mn^{III}, Co^{III} or Fe^{III} ions replace Al^{III} ions in an AlPO structure, a redox catalyst is produced, as described fully in later sections of this chapter. Isolated Ti^{IV} centres in a silica also function as Lewis acidic centres (see Section 4.6).

Introducing Catalytically Active Isolated Sites into Open-Structure Solids (Aluminophosphates, AlPOs)

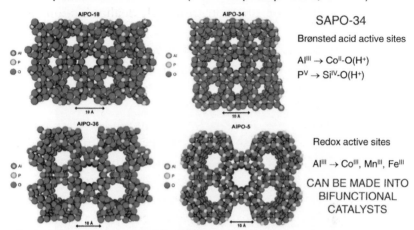

SAPO-34

Brønsted acid active sites

$Al^{III} \rightarrow Co^{II}\text{-}O(H^+)$
$P^V \rightarrow Si^{IV}\text{-}O(H^+)$

Redox active sites

$Al^{III} \rightarrow Co^{III}, Mn^{III}, Fe^{III}$

CAN BE MADE INTO BIFUNCTIONAL CATALYSTS

Three-dimensional representation of the pore structures of AlPO-18, AlPO-34, AlPO-36, and AlPO-5 (pore apertures are respectively 3.8 Å, 3.8 Å, 6.5 × 7.5 Å and 7.3 Å). In the AlPO-18 and AlPO-34 structures there are chabazitic cages as side pockets along the 3.8 Å diameter pores. (Note: the size of the atoms in the structures corresponds to their van der Waals radii.)

Figure 4.6 Four of the most widely used microporous aluminophosphates (AlPOs). By isomorphous substitution, as described later in the chapter, all these may be converted to Brønsted acid catalysts or redox catalysts or bifunctional ones — see Sections 4.4 to 4.9.

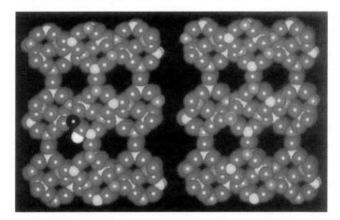

Figure 4.7 Computer graphic representation of the Co-containing AlPO-18 structure. The framework is very similar to that of the natural zeolite chabazite. The actual colours of the calcined and reduced forms of the catalyst are green and blue respectively. The blue catalyst has Brønsted acid centres; the green redox has centres as described in Sections 4.4 and 4.7 (red, oxygen; blue, Co^{III}; white, hydrogen; purple, phosphorus; yellow, aluminium).

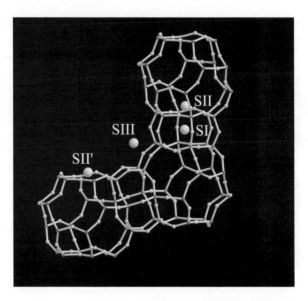

Figure 4.8 Position and nomenclature of the most important cation sites in X and Y zeolites (with the FAU structure). SI is in the so-called double-six ring (D6R). SII is in the sodalite cages, SII^I, is in the 6MR and SIII is within the supercage.[14]

As well as specific catalytic sites located in the framework of zeo-typic microporous solids, there are also extra-framework sites which often function as the principal source of catalytic activity. In faujasitic zeolites, for example, there are a number of well-identified sites that the extra-framework cations take up. The so-called SI, SII and SIII (sometimes designated S1, S2 and S3) sites are, along with others, shown in Figure 4.8. Whereas the SI (S1) site is at the centre of the double six-ring (D6R) that joins the two large cages, both SIII (S3) and SII (S2) are extra-framework sites, the SIII site being just above a four-membered ring (4MR), and the SII site above the 6MR, with each being in the supercage.

In situ techniques, such as parallel measurements of X-ray diffraction and X-ray absorption,[11] can reveal the movement of extra-framework cations, and other structural changes, during the course of catalytic turnover. Thus in the Ni^{2+}-ion-exchanged zeolite Y catalyst for the cyclotrimerization of acetylene,[12] it has been found that Ni ions originally present in the S1 site are extricated out into the S2 site (Figure 4.9), the driving force of the formation of the Ni-acetylene complex being more than adequate to overcome the barrier for the Ni ions to emerge from S1.[13] Extra-framework cations in SAPO acidic catalysts, such as the Cu^{2+}, H^+ SAPO catalyst known as STA-7[14] (see Figure 4.10), have been well studied because, like other Cu-exchanged SAPO materials,[15] they are efficient in the selective catalytic reduction of NO with NH_3. The cation positions in STA-7, determined via synchrotron radiation studies, are shown in Figure 4.10 along with an indication of the catalytic performance of the solid.[14]

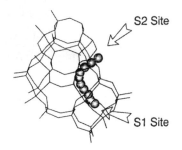

Figure 4.9   Migration pathway of the Ni cation from the S1 site to the more accessible S2 site. (The notation S1 and S2 is often used for SI and SII respectively.)[13]

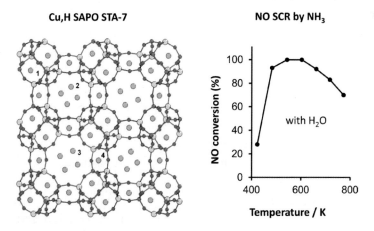

Figure 4.10   Projected structure of the active centre of Cu^{2+}, H^+ SAPO (STA-7) catalyst for the reduction of NO with NH_3. (After Wright.[14])

4.3 Some Examples of Acidic Microporous SSHCs

Industrially, in petrochemical and related processes, by far the most important and widely used microporous catalysts in this category are those derived from faujasitic zeolites. As outlined later, these provide an abundance of quasi-free protons that facilitate the breakdown of large hydrocarbon molecules — known as cracking, and as hydrocracking when accomplished in the presence of H_2. But very many acidic zeotypic catalysts are important both in industrial and academic contexts in such processes as isomerization, dehydration, esterification and alkylation. Figure 4.11 summarizes the essence of how a Brønsted acid catalyst functions. In addition to the alkylation shown here — where H-ZSM-5 is the catalyst of choice for the clean, selective production of ethylbenzene — one

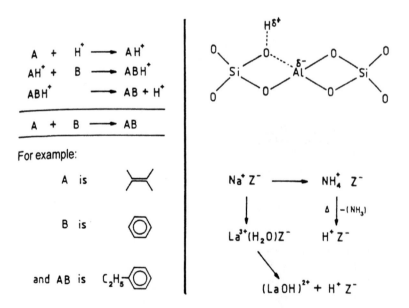

Figure 4.11 A Brønsted acid catalyst functions through its ability to donate protons (top left). There is one proton for every tetrahedrally bonded Al^{III} in the zeolite (top right). Zeolitic catalysts are often rendered acidic (H^+Z^-) by exchange with NH_4^+ ions followed by heating, or by simple exchange with polyvalent ions (e.g. La^{3+}) which then hydrolyse the bound water (bottom right).

may picture the simple isomerization of 1-butene to the more valuable 2-methylpropene as proceeding according to the sequence shown in Figure 4.12.[16] And as for the example of La[3+] zeolite Y, which is the backbone of the catalytic cracking industry, the root cause of the Brønsted acidity here is cation hydrolysis, which is also depicted in Figure 4.13, based on a neutron diffraction study[17] carried out to locate the La ion at a single site and the protons loosely bound to framework oxygen. (The protons become "visible" only at very low temperatures, *ca* 5 K, as, at higher temperatures, they are free to take up several equivalent positions on the oxygens of the framework.)

The question of whether carbenium and carbonium ions are implicated as reaction intermediates in zeolite-catalysed reactions has been discussed fully, and quantitatively using density functional

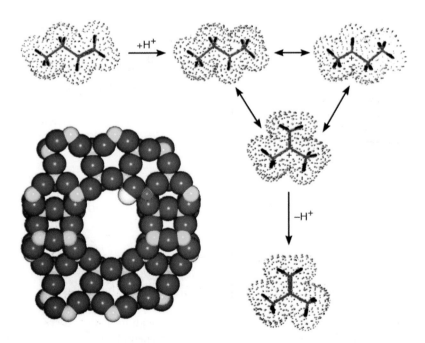

Figure 4.12 The acid-catalysed isomerization of 1-butene (top left) to 2-methyl-propene (bottom right) occurs rapidly on the inner walls (which provide protons) of microporous catalysts such as SAPOs and M^{II} AlPOs (M^{II} = Zn, Mg, Co, Mn).[16]

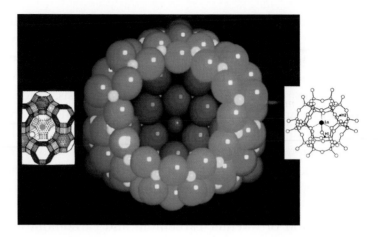

Figure 4.13　The La^{3+} ion in La-exchanged zeolite Y (faujasite) occupies a single site in the supercage. Owing to its strong polarizing power, a residual water molecule of hydration undergoes cation hydrolysis, thus generating "free" protons (for Brønsted acidity) (see text and Figure 4.11).

theory (DFT), by Boronat and Corma.[18] Such work builds on the pioneering endeavours of Haag and Olah, whose discussions of superacidity invoked five-coordinate carbon atoms. At an elementary level we may appreciate the pivotal role of acidic zeolites in processes such as cracking and hydrocracking from the following sequence of reactions:

$$>C=C< + H^+ \rightleftharpoons R_2^+$$

which describes how an alkylcarbenium ion is formed from alkenes, and

$$R_1-H + R_2^+ \rightleftharpoons R_1^+ + R_2-H$$

by hydride transfer. The alkylcarbenium ions can undergo three possible types of reaction:

rearrangement:　　　$R_1^+ \rightleftharpoons R_2^+$

β-cleavage:　　　　　$R_1^+ \rightleftharpoons R_2^+ + >C=C<$

addition to alkene:　$R_1^+ + >C=C< \rightleftharpoons R_2^+$.

(Acid catalysis of this kind was postulated by early workers such as Ipatieff and Pines whose catalysts were made by lacing siliceous solids — such as kieselguhr — with concentrated mineral acids.)

4.3.1 *Environmentally benign, solvent-free alkylations, acylations and nitrations using acidic SSHCs[19-22]*

All the processes to be described here are striking examples of shape-selective conversions using zeolitic acid catalysts. It will be seen that, for these purposes, H-ZSM-5 and H-Beta are the principal single-site catalysts employed.

Ethylbenzene (the raw material for styrene manufacture) is a classic example of a material produced by Friedel–Crafts alkylation, which is used extensively in both fine and bulk chemical industries. The original process involved AlCl₃ as a catalyst: it was corrosive and entailed the production of much acidic waste. In the 1980s, the so-called Mobil–Badger process, which simply entails the addition of ethylene to benzene over an H-ZSM-5 SSHC, supplanted it. Here, there is a stable, clean, recyclable catalyst, which has the added advantage of suppressing the formation (by shape selectivity) of polyalkylation products. A similar principle is involved in the Dow Chemical production of cumene (by the addition of propylene to benzene), the raw material for phenol production. Here, however, a 3D delaminated mordenite acid catalyst (designated 3-DDM) is used in place of H-ZSM-5. The dealumination increases the Si:Al ratio from 10 to 30 to *ca* 100 to 1,000, and, at the same time, changes the total pore volume and pore size distribution of the mordenite: the consequential presence of mesopores facilitates the diffusion of the reactants to the active sites. The Dow Chemical Company also pioneered the shape-selective dialkylation of polyaromatics, e.g. naphthalene and biphenyl (see Scheme 4.2) using the same type of 3-DDM catalyst. The products here are raw materials for the production of corresponding dicarboxylic acids, which are important industrial monomers for a variety of high-performance plastics and fibres.

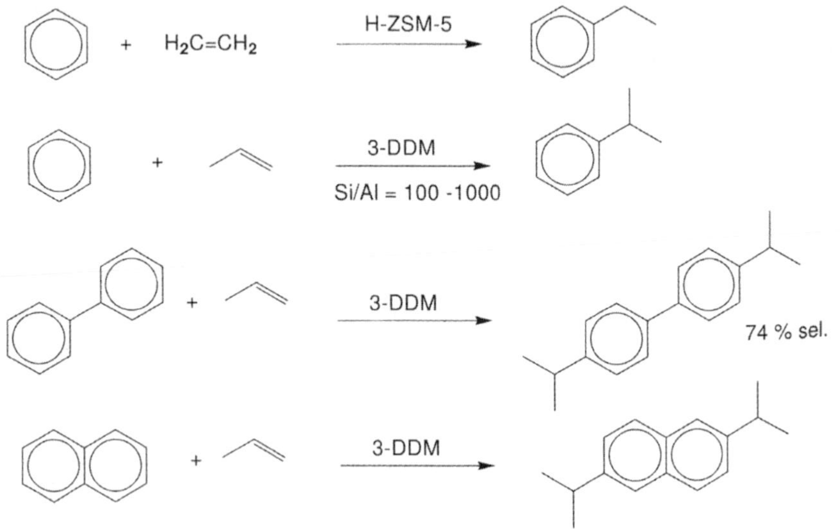

Scheme 4.2 A selection of zeolite-catalysed Friedel–Crafts alkylations. (After Sheldon *et al.*[22])

2,6-Dialkylnaphthalenes are readily oxidized to naphthalene-2,6-dicarboxylic acid that is used in the synthesis of the commercially valuable polymer, poly(ethylene naphthalenedicarboxylate) (PEN) — see Scheme 4.3. PEN has properties that are superior to those of PET (the analogous terephthalate), and is the polymer of choice for a variety of applications in industrial fibres, films, coatings, inks and adhesives. Smith and Roberts[21] optimized the dealuminated mordenite acid catalyst for this dialkylation.

For reasons similar to those described, Friedel–Crafts acylation using $AlCl_3$ (or $TiCl_4$) catalysts are nowadays environmentally unacceptable. Smith *et al.*[20] have developed an environmentally clean, solvent-free method of synthesizing a range of aromatic ketones by acylation of aryl ethers with carboxylic anhydrides over a zeolite Beta (H^+ form) at modest temperatures (*ca* 100 °C), the *para*-acylated products being obtained in high yield because of the operation of shape selectivity (see Figure 4.14).

Using the same acidic zeolite catalyst, Smith *et al.*[19] have evolved a clean (green) method of nitrating simple aromatic compounds

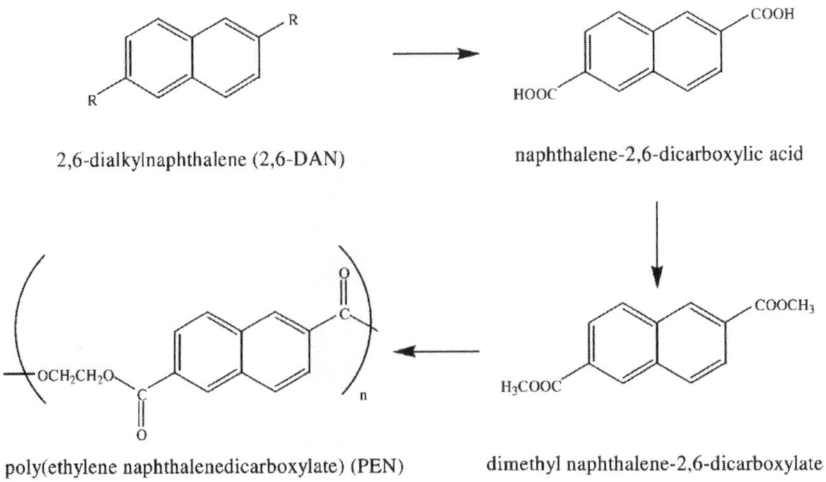

2,6-dialkylnaphthalene (2,6-DAN)

naphthalene-2,6-dicarboxylic acid

poly(ethylene naphthalenedicarboxylate) (PEN)

dimethyl naphthalene-2,6-dicarboxylate

Scheme 4.3 Synthesis of poly(ethylene naphthalenedicarboxylate) (PEN).[21]

such as benzene, alkylbenzenes, halogenobenzenes and some disubstituted benzenes. Again the operation of shape selectivity (Figure 4.14) generates very high yields of the *para*-substituted products. For example, nitration of

- toluene leads to quantitative mononitro products, of which 79% is 4-nitrotoluene;
- fluorobenzene also yields quantitative mononitro products, of which 94% is 4-nitrofluorobenzene; and
- 2-fluorotoluene gives 96% of mononitro products, of which 90% is the 5-nitro isomer and 10% the 4-nitro one.

4.3.2 *Brønsted acidic microporous SSHCs for hydroisomerization (dewaxing) of alkanes: designing new catalysts in silico*

Hydrotreating, which induces long-chain alkanes to isomerize as well as to break up into smaller and/or branched daughter molecules, is a major practice in the petrochemical industry. It relies crucially on having an adequate concentration (on the internal faces of microporous zeotypes) of Brønsted acid active sites (Figure 4.15). ZSM-5 and ZSM-11 (i.e. MFI and MEL) zeolites are

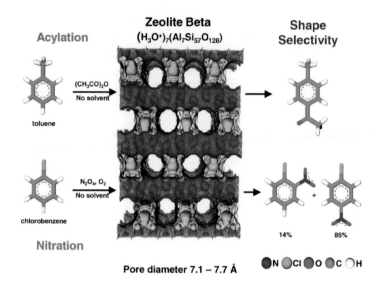

Figure 4.14 Shape selectivity (with acidic zeolite Beta) in the acylation of toluene and the nitration of chlorobenzene (see Smith *et al.*[19–21]).

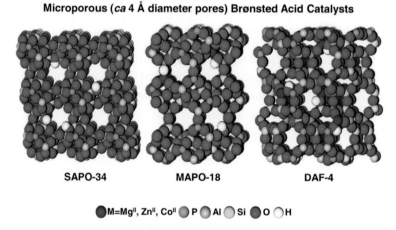

Figure 4.15 All three of these microporous, framework-substituted AlPOs exhibit high Brønsted acid catalytic activity in the dehydration of methanol and ethanol.

popular for this purpose; but there is a continual search for other, possibly superior ones. The synthetic experience of practitioners like Vaughan *et al.*[24] has brought forth new dewaxing catalysts, such

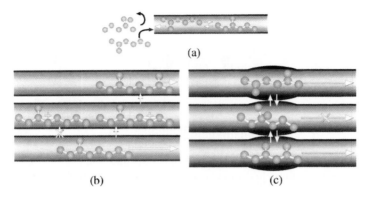

(a)

(b)   (c)

Figure 4.16 Illustration of the essential mechanisms giving rise to shape selectivity. (a) Reactant shape selectivity: molecules that are too large to enter the zeolite pores cannot reach acid sites for reaction and are therefore not converted into products. (b) Transition-state shape selectivity: molecules (and transition states) that are too large to fit inside a pore do not form. (c) Product shape selectivity: new molecules are formed in the adsorbed phase, but are too large to desorb as a product. (After Smit and Maesen.[26])

as the SAPO solid known as ECR-42. But such is the degree of theoretical-computational understanding of how hydroisomerization works[25,26] that it is now possible, as we outline here, to design new (as yet not experimentally synthesized) dewaxing catalysts.

At the heart of dewaxing, and hydroisomerization in general, is the concept and reality of shape selectivity which is illustrated here (see Figure 4.16)[26] and which manifests itself in three distinct ways:

- *reactant shape selectivity*: those reactant molecules that are too large to enter the pores of the catalyst cannot reach the acid sites for reaction to ensue;
- *transition-state shape selectivity*: molecules (and transition states) that are too large to fit inside a pore do not form; and
- *product shape selectivity*: new molecules may be formed in the adsorbed phase, but are too large to diffuse out as product.

At the Brønsted acid sites on the internal surface of a hydro-isomerization (dewaxing) microporous catalyst, a linear alkane is

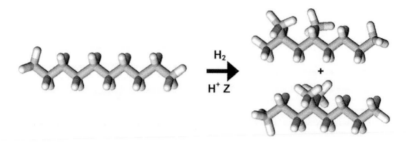

Figure 4.17   The hydroisomerization of linear alkanes, such as *n*-decane depicted here, to produce branched ones, is greatly facilitated by Brønsted acid zeolitic catalysts.

first converted to a protonated cyclopropane (carbocationic) intermediate that then proceeds to the product molecules, the nature of which is governed by the topography and tortuosity of the pores in the chosen catalyst. For example, when *n*-decane is catalytically hydrotreated, it may yield 2,4-dimethyloctane or 4,4-dimethyl-1-octene (along with smaller hydrocarbons) — see Figure 4.17. Such processes are involved in the dewaxing of (linear) hydrocarbons.[26] By inserting pendant methyl groups along the chains the alkane molecules do not crystallize as easily, hence suppressing the formation of waxy alkanes that are detrimental to their use as liquid fuels and lubricants (see Figure 4.18).

Efforts made previously to calculate thermodynamically the product distribution, such as that shown in Figure 4.17 based on free energy values of the isolated (pure) species, were unsuccessful because it is the free energy of the individual molecules *inside* the micropores of the acidic zeolite that matters. Thanks to recent advances in computer simulation, pursued by Smit and Maesen,[26] there is now far greater insight into the phenomenology of hydrotreating. By adopting the so-called configurational-bias Monte Carlo approach,[27] Smit and Maesen can now simulate the behaviour of large molecules inside micropores — a substantial advance from the early molecular dynamics calculations (for alkanes in faujasite[28] and xenon in silicalite[29] which Smit and I carried out over 20 years ago).

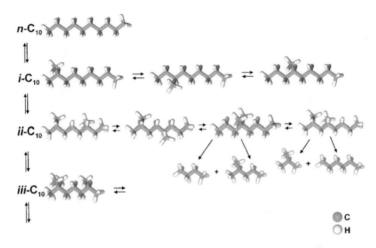

Figure 4.18 Some of the key reactions that take place over an acidic zeolite, such as those depicted in Figure 4.19.

So successful has the Smit approach proven that, as well as computing sorption isotherms that coincide with experimentally determined ones, he can now predict (and indeed has already patented!) the comparatively new zeolite framework structures known as GON and SFE (see Figure 4.19) as prime candidates for future industrial catalytic use.

The insight provided by Smit and Maesen[26] also reveal that for the closely related frameworks MFI (ZSM-5) and MEL (ZSM-11) there are marked differences in free energies for some specific reaction intermediates involved in hydro-dewaxing. Thus the reaction path of nC_{10} (decane) inside MFI proceeds as follows:

$$nC_{10} \rightarrow 5\text{--}MeC_9 \rightarrow 4,4\text{--}MeC_8$$

whereas inside MEL the dominant path is

$$nC_{10} \rightarrow 5\text{--}MeC_8 \rightarrow 2,4\text{--}MeC_8.$$

All this greatly assists the petrochemical catalyst practitioner.

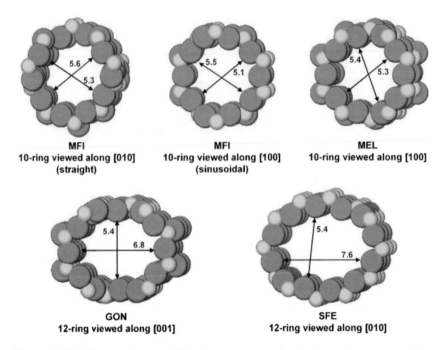

MFI
10-ring viewed along [010]
(straight)

MFI
10-ring viewed along [100]
(sinusoidal)

MEL
10-ring viewed along [100]

GON
12-ring viewed along [001]

SFE
12-ring viewed along [010]

Figure 4.19 The top two zeolitic (acidic) catalysts have framework structures designated MFI and MEL (i.e. ZSM-5 and ZSM-11, respectively) and are much used for shape-selective hydrotreating processes. Those known as GON and SFE, shown in the lower half, were identified *in silico* by Smit and Maesen[26] through their computational work. GON and SFE structures have recently been patented.

4.4 Brønsted Acidic Microporous SSHCs for the Dehydration of Alkanols: Environmentally Benign Routes to Ethylene, Propylene and Other Light Alkenes

This section deals first with the recent interest in producing ethylene from ethanol, a readily sustainable feedstock. We then consider the much more mature, but still somewhat enigmatic, process of converting methanol to light olefins — the so-called methanol-to-olefin (MTO) conversion.

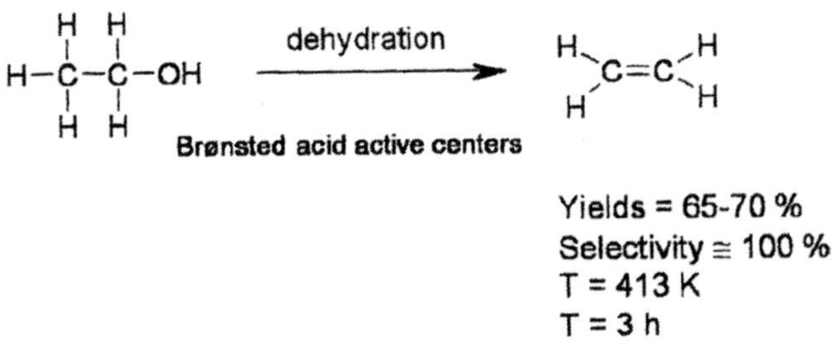

Figure 4.20 image showing:

$$H-\underset{\underset{H}{|}}{\overset{\overset{H}{|}}{C}}-\underset{\underset{H}{|}}{\overset{\overset{H}{|}}{C}}-OH \xrightarrow[\text{Brønsted acid active centers}]{\text{dehydration}} \underset{H}{\overset{H}{\diagdown}}C=C\underset{\diagdown H}{\diagup H}$$

Yields = 65-70 %
Selectivity ≅ 100 %
T = 413 K
T = 3 h

Figure 4.20 A new range of open-structure Brønsted acid catalysts show great promise for the conversion of ethanol to ethylene.[30,31]

4.4.1 Catalytic dehydration of ethanol using Brønsted acidic SSHCs

The well-known undergraduate exercise involving concentrated sulfuric acid as a dehydrating agent for the conversion of ethanol to ethylene can be carried out much more cleanly and smoothly using narrow-pore solid acid catalysts, such as those shown in Figure 4.15 at moderate temperatures. With there being no shortage of ethanol (from bio-sources) at present this single-site catalytic reaction (see Figure 4.20) is a viable method of producing the olefin feedstocks (ethylene particularly) required for industrial production. Using a range of MAPO-5 (M = Mg^{II}, Zn^{II}) as well as SAPO-5 catalysts, Paterson *et al.* have reported conversions of ethanol as high as 80% and a selectivity towards ethylene of over 95% (with TOFs of *ca* 900 h^{-1}).[30,31]

4.4.2 The methanol-to-olefin conversion over Brønsted acidic SSHCs

Shortly after the Union Carbide Co. (as it was then called) announced in 1982 the discovery of the extensive family of AlPO and SAPO zeotypic solids, reports from pioneers such as Flanigen and Rabo indicated that, whereas the pentasil solid acid catalysts ZSM-5 and ZSM-11 were efficient in converting methanol to gasoline, the narrower-pore SAPO-34 (see Figure 4.15) produced only light olefins (C_2, C_3 and C_4), the supposed rationale for this observation being that the channels in SAPO-34 are too narrow to allow for the formation (owing to shape selectivity) of the aromatic molecules benzene, toluene and xylene. This report prompted several subsequent studies. In an early one[32] it was shown that, in addition to the light olefins, light alkanes were also generated.

The MTO (MTH) — methanol-to-olefin (methanol-to-hydrocarbon) — reaction is now of enormous commercial significance worldwide. Owing to depletion of oil reserves and increasing oil prices, interest in the MTO process is as high as ever, and industrial-scale methanol-to-propene (MTP), methanol-to-gasoline (MTG) and MTO plants are currently in operation or under construction in China, all based on coal as raw material.[33] (The CO/H_2 mixtures generated by the controlled combustion of coal may be readily converted to methanol by the ICI $Cu/ZnO/Al_2O_3$ or other such catalysts.) Moreover, the successful start-up of a demonstration plant (2010) for the MTO process based on natural gas in Belgium was recently reported, and a demonstration plant for the MTG process based on biomass is currently under construction (by the Topsøe Company) in the US (Figure 4.21).

As well as SAPO-34, it was shown by Chen and Thomas[34] that MAPO-18 (with M = Co^{II}, Mg^{II}, Mn^{II}, Zn^{II}) is very efficient in producing both ethylene and propylene, with traces of butene under mild conditions of Brønsted acid catalytic dehydration of methanol. There have since been reams of studies reporting the MTO reaction, but its complexity remains rather elusive. Arguments still rage as to the mechanism responsible for the formation of the first C-C bond once the methanol has been converted (as has been established experimentally) to the dimethyl ether. This is an auto-catalytic reaction, in which the products react faster with methanol than methanol itself.[35] In 1994, after numerous experiments had been carried out, Dahl and Kolboe[36] introduced the concept of a "hydrocarbon pool", implying that a reservoir of hydrocarbons bound to the inner surface of the zeolitic catalyst remains in the framework to create a special active site, consisting of the Brønsted acidic inorganic structure juxtaposed to an organic carbocationic molecule. Figure 4.22 taken from the work of Haw *et al.*[37] (who used *in situ* solid-state NMR) represents one plausible picture of the nature of the active catalyst. It is as if the inorganic acidic catalyst

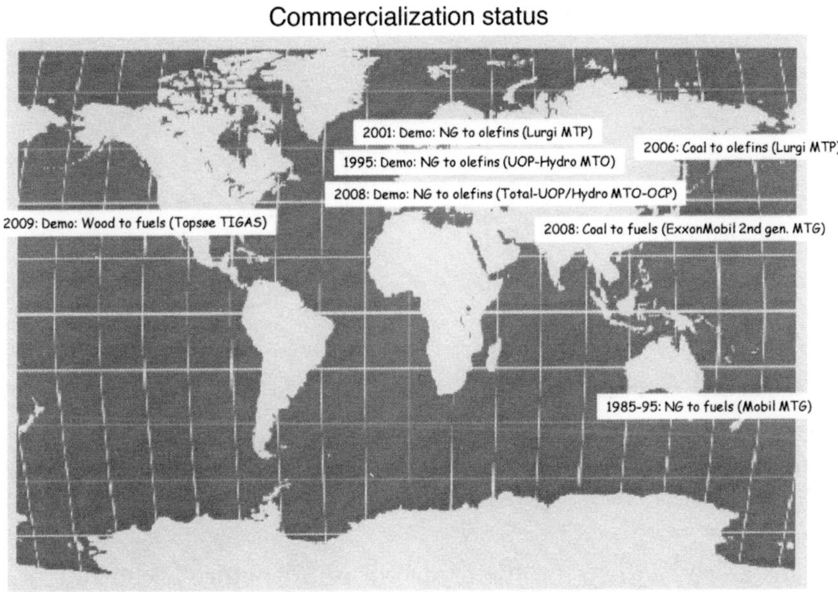

Commercialization status

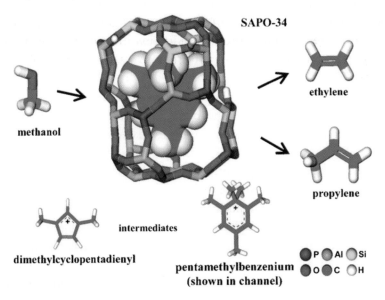

2001: Demo: NG to olefins (Lurgi MTP)

1995: Demo: NG to olefins (UOP-Hydro MTO)

2006: Coal to olefins (Lurgi MTP)

2008: Demo: NG to olefins (Total-UOP/Hydro MTO-OCP)

2009: Demo: Wood to fuels (Topsøe TIGAS)

2008: Coal to fuels (ExxonMobil 2nd gen. MTG)

1985-95: NG to fuels (Mobil MTG)

Figure 4.21 Manufacturing and demonstration plants for the methanol-to-olefin (MTO) process are distributed over the world, as this map, by U. Olsbye of the University of Oslo, shows.

SAPO-34

methanol

ethylene

propylene

intermediates

dimethylcyclopentadienyl

pentamethylbenzenium
(shown in channel)

P   Al   Si
O   C   H

Figure 4.22 Haw and others[37] have identified cationic hydrocarbon intermediates, some of which are shown here, inside the SAPO-34 Brønsted acid catalyst.

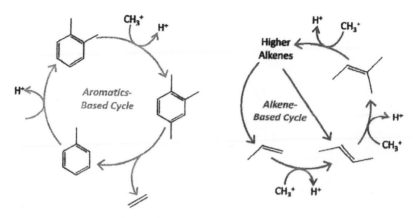

Figure 4.23 The aromatics-based (left) and alkene-based (right) reaction cycles. (After Olsbye *et al.*[33])

needs a "co-factor" (the hydrocarbon pool) before it can function to generate the olefins.

The complexity of the MTH reaction mechanism continues to intrigue the academic community. A significant breakthrough in mechanistic understanding came when Svelle *et al.*[38] used transient isotopic labelling experiments over the 3D ten-ring MFI (ZSM-5) topology. These workers proposed that two inter-related, main reaction schemes run in parallel within the nanopores of the 3D MFI structure:

- an alkene-based methylation and cracking route leading predominantly to propene and higher alkenes; and
- an aromatics-based methylation and dealkylation route leading predominantly to ethene (see Figure 4.23).

This dual-cycle mechanism proposal departed from the then prevailing picture that methyl benzene intermediates (such as those shown in Figure 4.22) were essential for the formation of alkenes from methanol. This insight led, in turn, to the prediction that a ten-ring 1D topology might favour the alkene-based route to a larger extent, thereby yielding C_3^+ alkene formation and suppressing the

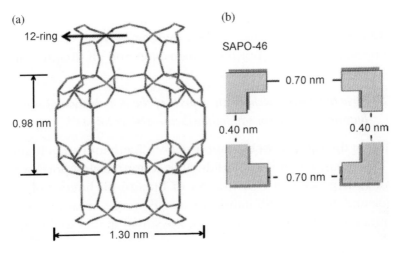

Figure 4.24 (a) The cage and pore structure of SAPO-46; and (b) the cage structure in the framework. (After Dai *et al.*[40])

production of aromatics. This hypothesis was verified by an experimental study[33] of the 1D ten-ring (Theta-1) TON topology, which gave more than 70% selectivity to branched C_5^+ (gasoline-fraction) alkenes and less than 1% to aromatics at 400 °C. The key message here, therefore, is that the precise performance of the Brønsted acidic active (single) site is mediated by the pore dimensionality of the solid catalyst. Even when the pore *dimensions* of two nanoporous SSHCs are similar, the pore *dimensionality* (i.e. whether 1D or 3D) holds sway.

Fundamentally important new items of experimental information continue to come to light considering the MTO reaction of acidic SSHCs. In December 2010, Dai *et al.*[40,41] demonstrated that detrital remnants of the organic template (used to synthesize the SAPO catalyst) play a decisive role in the MTO reaction. For the zeotype SAPO-46 (see Figure 4.24) the organic impurities toluene, benzyl alcohol and some higher alkanes are unmistakably present in calcined samples of the acidic catalyst. It was shown by these workers that toluene, in particular, may be the primordial hydrocarbon pool or the precursor for the formation of the primordial hydrocarbon pool that induces the initial MTO reaction to take place. Under the MTO reaction conditions, residual toluene in the SAPO-46 catalyst can be transformed to polymethylbenzenes and alkylnaphthalenes.

4.4.3 *Structural and mechanistic aspects of the dehydration of isomeric butanols over porous aluminosilicate acid catalysts*[42]

Whereas the catalytic dehydration of both methanol and ethanol, as we have seen, are of very considerable commercial importance, the analogous process of dehydration of butan-1-ol, butan-2-ol, isobutanol and tertiary (*t*)-butanol are rather more of academic than commercial interest at present. But the situation may well change as higher olefins (formed by oligomerization of lighter ones generated in the initial act of dehydration) could well become industrially more significant than they are at present.

The four butanols constitute an ideal set of molecules in which to assess the relative importance of structure (of the SSHC) on the one hand and the mechanistic details of dehydration on the other. These four alkanols were the subject of detailed study by the present author, the late academician Zamaraev and co-workers in the early 1990s.[43–48]

Seldom in the study of heterogeneous catalysts does it prove possible to (i) specify precisely the concentration and nature of the active sites, (ii) test whether these sites are of comparable strength and are distributed in a spatially and chemically well-defined manner and (iii) explore the structural and mechanistic features of the system using a wide range of complementary techniques, many of them *in situ*. Even rarer are situations in which both the access to the active sites and the shape of the reactants may be systematically and subtly varied, so that one is able to compare the performance of the active site in a crystalline (microporous) environment with an essentially identical one embedded in an amorphous solid.

The techniques used (by Zamaraev and Thomas *et al.*) included solid-state NMR with ^1H, ^2H, ^{13}C, ^{27}Al and ^{29}Si nuclei, *in situ* FTIR, chromatography and isotopic labelling as well as kinetic studies. The microporous solid acid catalyst was H-ZSM-5 of several different

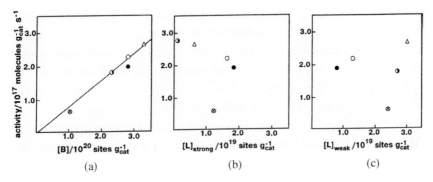

Figure 4.25 Relation between the activity of ZSM-5 and amorphous aluminosilicate (AAS) in the catalytic dehydration of isobutanol to butane at 397 K and the number of (a) Brønsted acid sites [B]; (b) strong Lewis acid sites $[L]_{strong}$; and (c) weak Lewis acid sites $[L]_{weak}$. (Δ) NaH-ZSM-5 sample; (●) H-ZSM-5 sample; (○) another H-ZSM-5 sample; (●) yet another H-ZSM-5 sample with yet another different concentration of sites; (⊗) AAS. (After Zamaraev and Thomas.[42])

crystallite sizes, and with Si/Al ratios ranging from 20 to 35 to 42. The concentration of the active sites, in units of 10^{20} g^{-1}, ranged from 2.3 to 3.3. The amorphous solid acid catalyst — a silica-alumina preparation — had a pore diameter of *ca* 50 Å (compared with *ca* 5.5 Å for the ZSM-5 catalysts) and its crystallite size was *ca* 1,000 μm. H-ZSM-5 samples ranging in crystallite sizes from < 1 μm to 4–6 μm to 15–20 μm were used. The (Brønsted) active site in all the samples is the same, namely ≡ Al-O(H)-Si ≡.

It is seen from Figure 4.25 that dehydration is catalysed by the Brønsted acid, not by the strong or by the weak Lewis acid sites (which can be independently determined by adsorption of pyridine). This figure also shows (from the linearity of the plot) that the catalytic activity per unit site is the same for *all* of the acid catalysts under study.

Clearly, the activity of all the catalysts is directly proportional to the concentration of the bridging hydroxyl groupings and does not correlate with the concentration of strong or weak Lewis acid sites. Further proof that Lewis acid sites are not catalytically active, whereas Brønsted ones are, comes from the fact that preadsorbed acetonitrile (which affects Lewis but not Brønsted sites) has no effect on the rate of dehydration, whereas increasing amounts of preadsorbed pyridine (which becomes attached to Brønsted sites) progressively decreases the rate of dehydration (of isobutanol). The concentration of ≡ Al-O(H)-Si ≡ active sites was measured by four distinct methods: ^{1}H NMR spectroscopy; quantitative IR spectroscopy; "poisoning" titration with pyridine; and direct analysis of the Al content. Each of these was in good agreement with one another.

From the solid-state ^{2}H NMR spectra of *t*-butanol adsorbed within samples of H-ZSM-5 catalysts, it was concluded that the geometry of the isolated molecule is unperturbed when bound to the active site. Since the critical dimension of the *t*-butanol (*ca* 6.8 Å) significantly exceeds the channel diameter of the catalyst (*ca* 5.5 Å) — see Figure 4.26 — it follows that this reactant molecule must be accommodated at channel intersections, where the free diameter is *ca* 9 Å. Therefore we conclude that the Brønsted active sites are themselves located at such intersections.

We deduce that the Brønsted acid active sites in the H-ZSM-5 samples and the amorphous aluminosilicate sample are of the same acid strength because the overall shift in OH stretching (IR) frequency when CO is adsorbed at low temperature — a well-known test — is essentially the same. (The magnitude of this shift is directly proportional to the Brønsted acid strength.[42])

The kinetics of adsorption and dehydration of the four butanols were measured *in situ* via the time dependencies of the (IR) line intensities at 1,460–1,470 cm^{-1} (which reflect the CH deformation vibrations) and 1,640 cm^{-1} (the deformation vibrations of adsorbed H$_2$O), respectively. It was found that adsorption of the *n-*, *sec-* and isobutanols was complete within 25 s at 296 K. The dehydration process of these alcohols in the zeolitic pores was, however, slower. And for a given alcohol (*n-*, *sec-* or iso-) the kinetics of water elimination were identical for catalysts of different crystallite sizes. This established the absence of any diffusion limitation for dehydration for these three alcohols.

For the bulkier *t*-butanol, the situation was different. The kinetics of its adsorption at room temperature were markedly retarded

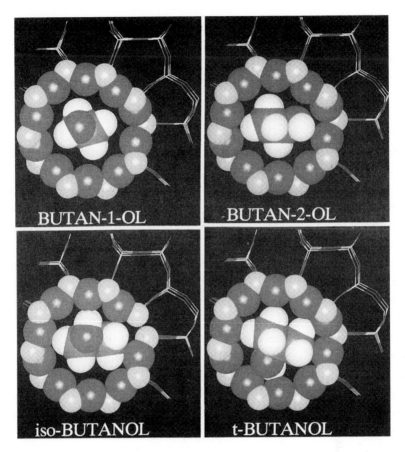

Figure 4.26 Graphic illustration of the cross-sectional views of the four isomeric butanols, studied in Zamaraev and Thomas,[42] in relation to the 5.5 Å diameter of the ZSM-5 pore.

with increasing crystallite sizes of the H-ZSM-5; and, moreover, they exhibited a $t^{1/2}$ dependence (t is the time after adsorption). This is symptomatic of a diffusion-influenced process. Indeed it was possible to measure the diffusion coefficient of t-butanol in H-ZSM-5 at 296 K as 2×10^{-11} cm^2 s^{-1}.

For all four butanols in the zeolitic catalysts with small enough crystallite sizes — when diffusion limitations also disappear — dehydration kinetics are well expressed by the Arrhenius exponential function, a fact that is explicable in terms of the unimolecular decay of molecules of the butanols adsorbed on identical active sites. With isobutanol, for example, the rate coefficient k may be written

$$k = 2 \times 10^9 \exp\left(\frac{-80 \pm 10kJ}{RT} mol^{-1}\right)$$

where the rather small pre-exponential factor corresponds to an activation entropy of -70 kJ mol^{-1} K^{-1}.

All the steady-state and transient kinetic data for the dehydration of all four butanols using H-ZSM-5 and the amorphous aluminosilicate catalysts were rationalized by Zamaraev and Thomas as shown in Figure 4.27.

This one, all-embracing scheme (shown in Figure 4.27) satisfactorily accounts for the reaction mechanism of all four butanols over the Brønsted acid (SSH) catalysts. Reaction pathways were found[42] to be identical for all the butyl alcohols and the catalysts studied by Zamaraev and Thomas *et al.*, both with respect to the main reaction stream and to the side reactions. However, depending upon the

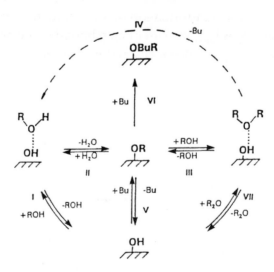

Figure 4.27 Reaction mechanism that rationalizes the catalytic dehydration processes (see text).

particular butanol, the observed reaction rates and selectivity towards various products can be significantly varied.

Three main conclusions can be drawn from this work. The first refers to the nature of the key reaction intermediate $\overset{OC_4H_9}{\underset{IIII}{-I-}}$, which clearly exhibits carbenium-ion properties such as scrambling of carbon and hydrogen (isotopically labelled) over its skeleton. In this respect, the heterogeneous dehydration of the butanols over the solid acid catalysts mechanistically resembles homogeneous acid-catalysed reactions, whereby carbenium ions in their classical forms serve as key reaction intermediates.[49] But the acid-bound -OC_4H_9 intermediate does not coincide exactly with the classical carbenium ion for the following reason: the -OC_4H_9 surface intermediate can exist in three different states, namely (i) in an ion pair involving a butyl carbenium ion and the $\equiv$ Al-O(H)-Si $\equiv$ of the catalyst, (ii) interconversions between these states (as depicted in Scheme 4.4) and (iii) intramolecular rearrangements in the butyl carbenium ion $\equiv$ Al-O(H)-Si $\equiv$ ion pair, proceed with finite rates that vary with the temperatures.

It is clear that the reaction intermediate -OC_4H_9 for the dehydration of butanols can exist as butyl silyl ether (BSE) and adsorbed butyl carbenium ion (BCI). And NMR measurements and kinetic data show that there is reversible transformation between BSE, BCI and adsorbed butene (denoted by Bu_{ads}), the latter being bound to the bridging acid site as ethylene and other alkenes as in the work of Zecchina *et al.*[50]

The manner in which an isotopically labelled carbon atom is scrambled within an adsorbed butyl carbenium ion (IBCP) and isobutyl silyl ether (IBSE) in the case of isobutanol is shown in

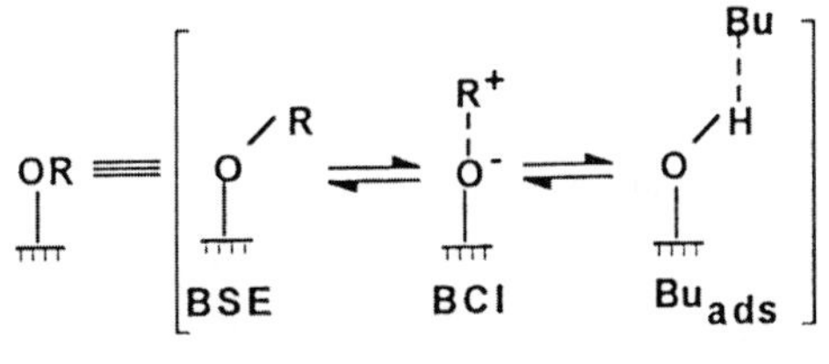

Scheme 4.4 Summarized picture of the catalytic dehydration of bound OR (where R = C_4H_9). BSE stands for butyl silyl ether, BCI for butyl carbenium ion and Bu_{ads} for adsorbed butene.[42]

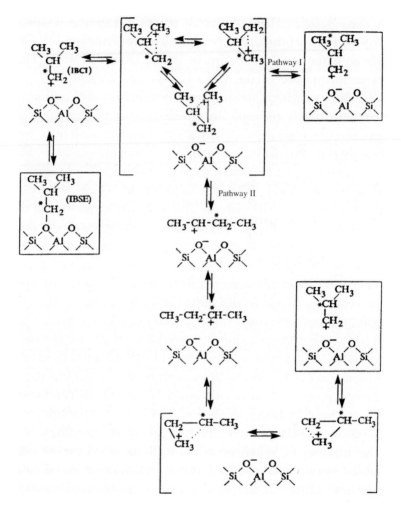

Figure 4.28 The inter-relationship between isobutyl carbenium ion (IBCI), isobutyl silyl ether (IBSE) and the various pathways shown in Figure 4.27.[42]

Figure 4.28. Interaction of a carbenium-ion state with the wall of the microporous catalyst can kinetically favour reactions that are not favoured for carbenium ions in solution. In particular, pore confinement in H-ZSM-5 favours the formation of linear $C_4H_9^+$, although branched $C_4H_9^+$ ions are favoured in solution both kinetically and thermodynamically.

4.5 Lewis Acidic Microporous SSHCs for a Range of Selective Oxidations

When the Italian workers (Taramasso *et al.*[51]) at the EniChem Company in Milan announced in 1983 that single-site Ti^{IV} centres inside a silicalite (MFI) framework could function as powerful Lewis catalysts for a range of selective oxidations (see Scheme 4.5), the news was greeted with considerable scepticism. This was because Pauling's rules, as well as prior practical experience, had indicated that Ti^{IV} ions could not take up tetrahedral coordination. An abundance of evidence, subsequently garnered,[52–54] has provided inexpugnable proof that Ti^{IV} ions do indeed take up tetrahedral sites, not only in MFI, but in zeolite Beta[55] and in mordenite (as well as in mesoporous silicas, which we shall consider more fully in Chapter 6). It was also demonstrated that tetrahedrally coordinated Ti^{IV} ions could be accommodated in isolated fashion

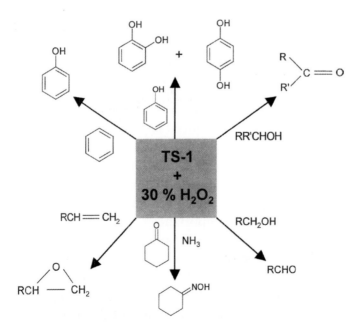

Scheme 4.5 Schematic representation of the most relevant oxidation reactions catalysed by TS-1. (After Zecchina *et al.*[53])

into SAPO-5[56] as well as AlPO-5 and AlPO-11.[57] Single-site Ti^{IV} ions have also been introduced to the microporous silicas ITQ-7[58] and MWW.[59,60]

Other metal ions that impart Lewis acidity have also been successfully introduced into the zeolitic MFI framework, notably Sn and Fe.[61] Moreover, the zeolite Beta framework (with its wider pores than MFI, thus being better suited for the selective oxidation of bulkier organic molecules) can also readily accommodate isolated Sn and Zr as well as isolated Ti ions.[61]

Many of these new SSHCs are ideally suited to process large (e.g. sugar) molecules and are already taking on a significant role in the realm of sustainable chemistry, where the feedstocks are from plant and other biological sources.[61–63] We shall return to the discussion of such topics, especially the conversion of sugars to lactic acid derivatives, in Chapter 5. Several of these Lewis acid microporous solids are particularly good in functioning as SSHCs for cascade reactions, a typical example of which, involving Ti^{IV}-substituted AlPO-5 (i.e. TAPO-5), we illustrate next.

4.6 Cascade Reactions with TAPO-5

When a Ti^{IV} ion is incorporated into the framework of AlPO-5 (see Figure 4.29) it confers not only Lewis acidity but weak Brønsted acid activity to the catalyst (as evidenced by the fact that the catalyst converts methanol to dimethyl ether and also releases nascent oxygen from H_2O_2). This bifunctionality — about which we shall say more later in this chapter and in Chapter 5 — makes it possible to carry out a cascade (a sequence) of catalysed reactions in "one pot". One-pot reactions are of increasing importance[63–66] because they constitute an efficient means of carrying out organic syntheses (in particular) and other useful transformations. It is estimated that, at present, about 80% of the cost of most chemical processes arises from the separation of products from the reaction mixture. In multi-step processes (which are invariably the practice in the pharmaceutical industry and also a central feature in the synthesis of natural products) the costs involved in successive separations are frequently extortionate.

TAPO-5

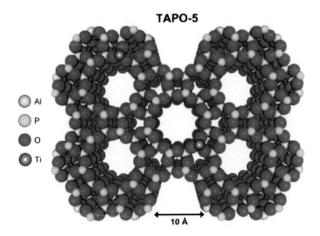

Al
P
O
Ti

10 Å

Figure 4.29 The structure of the multifunctional TAPO-5 catalyst where the four-coordinate framework-substituted Ti^{IV} ions catalyse all the cascade of reactions shown in Scheme 4.6.[65]

A recent laboratory-scale example of cascade reactions entails the conversion of cyclohexene by H$_2$O$_2$ to adipic acid. (*c*-Hexene is now readily produced industrially by the selective hydrogenation of benzene.) Ideally, it is better for O$_2$ rather than H$_2$O$_2$ to be used to prepare adipic acid, since four molecules of H$_2$O$_2$ are required per molecule of the alkene converted. We shall show later — see Section 4.7 — that these redox active sites in microporous solids do, indeed, convert both *n*-hexane and *c*-hexane into adipic acid at such single sites.

From the results of parallel ^{13}C and ^{1}H NMR and GC-MS analysis conducted during the course of the catalytic conversion of **1** to **2** in the presence of TAPO-5, and from additional analytical measurements starting with certain intermediates (notably **3** and **4**), we conclude that the mechanistic pathway from **1** to **2** is as shown in Scheme 4.6. Importantly, we found that this *cis*-diol (*cis*-**4**), but not the *trans*-diol (*trans*-**4**), is formed by a free radical mechanism.

This synthesis of adipic acid is solvent-free. Although the yield of the acid is not as high as in other methods of synthesis described later, we have uncovered here mechanistic details not hitherto detected and also established the key importance of the

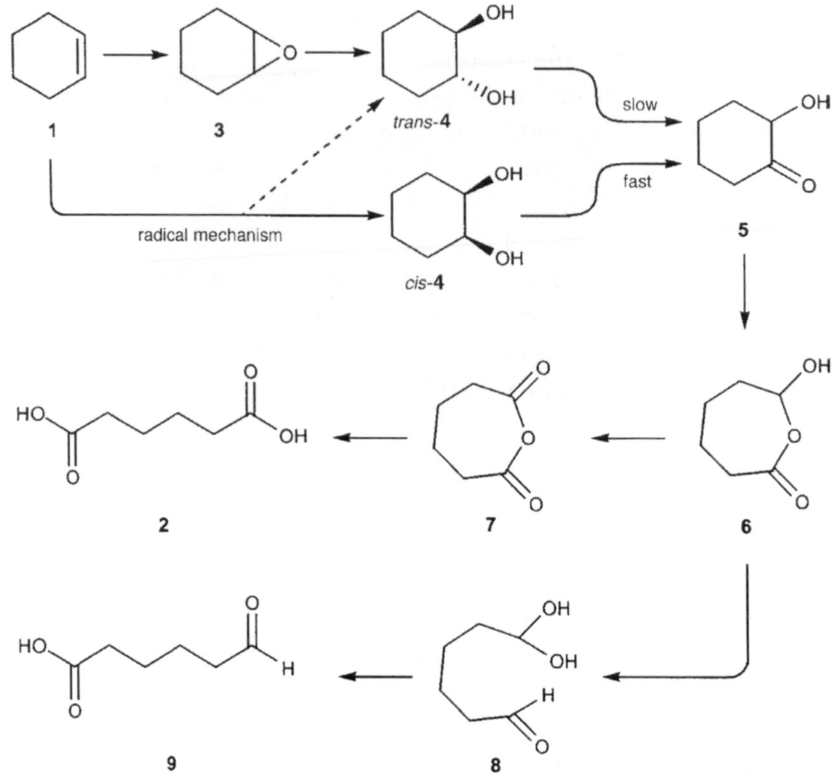

Scheme 4.6   The cascade of reactions involved in the conversion of cyclohexene to adipic acid using H_2O_2 as oxidant over a TAPO-5 catalyst.

stereochemistry of the 1,2-cyclohexanedione in the progress of the reaction. We return to this synthesis in Section 6.2.

4.6.1   *One-pot reactions: a contribution to environmental protection using Lewis acid active sites*

An extra reason for pursuing the development of cascade (one-pot) conversions is the avoidance of side product formation and loss of starting material, in addition to the reduction of operation costs, alluded to earlier. A particularly interesting example of using Lewis acid active sites for such a one-pot process is contained in the work of Hoelderich,[67] who produced phenol from benzene in high yields using a coordinatively unsaturated, extra-framework Al site as the locus of the benzene plus nitrous oxide reactant. (This is rather akin to the famous work of Panov *et al.*[68] using Fe-ZSM-5 and N_2O as the reactant for phenol production from benzene.)

By deliberately "extracting" in a gentle manner, using the well-known steam process for treating zeolites, Hoelderich significantly boosted the number of Lewis acid sites in the H-ZSM-5 sample (Si/Al *ca* 56) that he began with — see Figure 4.30. The well-known FTIR signatures of adsorbed pyridine show a substantial increase in Lewis acid, i.e. extra-framework, coordinatively unsaturated Al centres in the sample. It transpired that, by steaming the zeolite for 3 h or so, the Lewis acid active centres reached a maximum concentration. Hoelderich[67] found a direct correlation between the degree of extra-framework Al site concentration and the conversion of benzene to phenol by N_2O. (This author also reported[67] that picoline may be directly oxidized in O_2 in the gas phase to nicotinic acid in the presence of a V_2O_5-impregnated TiO_2 catalyst. This particular conversion, as we shall see in Chapter 5, may be smoothly carried out on SSHCs based on AlPO structures.)

Examples of other cascade (one-pot) reactions are given in subsequent sections of this chapter.

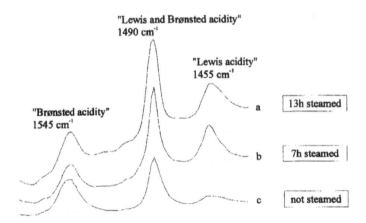

Figure 4.30 FTIR investigation of steamed H-[Al]-ZSM-5 catalysts showing the different kinds of acid sites. (After Hoelderich.[67])

4.7 Redox Active Sites in Microporous Solids

4.7.1 *Introduction*

In the early 1990s the author and his colleagues explored the ease with which redox centres (in addition to Brønsted and Lewis acid centres) could be introduced into various MAPO structures, with M = Co, Mn and Fe.[69–70] By redox centres we mean, typically, sites where one or other of these three transition metal ions may occur, depending upon the preparative conditions, as either M^{II} or M^{III} ions. When M^{II} ions occupy Al^{III} sites in the MAPO, an adjacent, loosely bound proton is also present as expected in the classic Brønsted acid, hydrogen-bridging case. The intention in "placing" M^{III} sites into the AlPO framework was partly to explore ways in which such redox AlPOs could be designed to oxidize regioselectivity or shape-selectivity alkanes in air or O_2 at low temperatures. Because AlPOs, which are quite thermally stable, have large internal surface areas (up to $1,000 \ m^2 g^{-1}$) and have an open structure, they offer ideal scope for delicate manipulation of catalytic properties.

There were two other considerations that prompted the author to explore the selective, oxidative power of Mn^{III}-, Co^{III}- (and Fe^{III}-) framework-substituted molecular-sieve catalysts, and both of these arose more because of poetic suggestion rather than indisputable scientific fact. First, Mn^{III} and Co^{III} ions accommodated substitutionally in place of Al^{III} ions in the MAPO framework possess close to tetrahedral coordination[71] and, just like four-coordinate Ti^{IV} ions in TS-1 or the five-coordinate Mn ions in manganese superoxide dismutase,[72] they are coordinatively unsaturated, a feature that facilitates redox catalytic reaction.

Second, the occurrence of isolated Mn^{III} (or Co^{III}) ions in the framework of the MAPO would, in the presence of oxygen and those alkanes whose size and shape permit access to these high-oxidation-state ions, favour the production of free radicals, which one knew[73,74] to be implicated in the selective oxidation of saturated hydrocarbons with oxygen.

Given that an enormous range of thermally stable (MAPO) solids are readily preparable, and that the local environment of the introduced element can be straightforwardly determined by X-ray absorption and other spectroscopic techniques,[16] they offer great scope for the development of new SSHCs.

4.7.2 *Single-site redox active centres for the benign selective oxidation of hydrocarbons in air or O_2[75]*

We now cite several examples of microporous (predominantly AlPO-based) solids with isolated redox centres that are efficient in commercially important selective oxidation. All these highlight the remarkable activity of Mn^{III} (as well as Co^{III} and Fe^{III}) catalytic centres for the activation of C–H bonds in alkanes and alkyl aromatics.

4.7.2.1 *The selective oxidation of cyclohexane in air by single-site molecular-sieve (redox) catalysts: towards a cleaner method of producing adipic acid[76,77]*

The incentive for designing a redox SSHC for oxidizing cyclohexane preferentially to adipic acid was to circumvent (on a laboratory scale) the environmentally aggressive and otherwise complex "nitric acid" procedure (used commercially — see Figure 4.31). Apart from its multi-step character and high costs this method produces the greenhouse gas N_2O as a by-product in large quantities.

Based on earlier work[16,75] on transition-metal-substituted AlPO-5, it was argued that it was necessary to "place" the Fe^{III} redox centre into a microporous structure, the pore aperture of which was significantly smaller than that of Fe AlPO-5 (7.3 Å diameter). The host AlPO that we selected was that with the so-called 31 structure, hence Fe^{III} AlPO-31 was chosen (see Figure 4.32) because it would constrain the intermediates that are known to be formed when cyclohexane is exposed to O_2 or air, *viz* the two free radicals $C_6H_{11}\cdot$ and $C_6H_{11}OO\cdot$ as well as the cyclohexyl hydroperoxide $C_6H_{11}OOH$, cyclohexanol and cyclohexanone on the path to the final product, adipic acid.

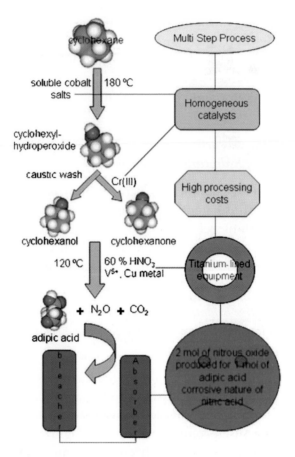

Figure 4.31 The "nitric acid" method for the production of adipic acid from fossil fuels liberates large quantities of the greenhouse gas N_2O, besides involving a complex combination of homogeneous and heterogeneous processes. (After Thomas *et al.*, *Chem. Eur. J.*, **7**, 2973 (2001).)

Product shape selectivity (see Section 4.3.2) is what one capitalizes upon here. The intermediates and free radicals enumerated earlier all diffuse out rather readily from the Fe^{III} AlPO-5 structure, so much so that the residence time for the cyclohexanol and cyclohexanone is too short for further conversion to adipic acid to occur within the pores. Since the Fe^{III} AlPO-31 has significantly narrower pores (5.4 Å), adequate time is available for the -ol and

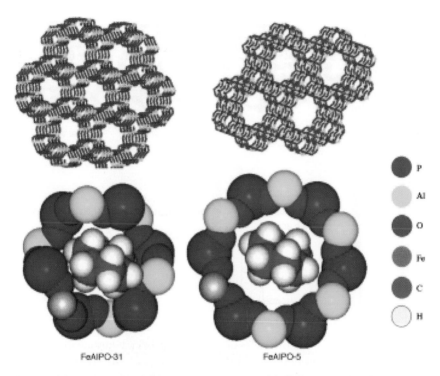

FeAlPO-31   FeAlPO-5

P
Al
O
Fe
C
H

Figure 4.32 (Top) A side view of microporous AlPO-31 and AlPO-5 structures, which show the one-dimensional channels. (Bottom) Representation of the cyclohexane molecule in the differently sized pore openings of Fe AlPO-31 (5.4 Å) and Fe AlPO-5 (7.3 Å).[76]

-one intermediates to convert to adipic acid, which, because of its shape — it is a more sinuous molecule than either the -ol or -one (which is globular) — diffuses out of the nanoporous catalyst freely, thereby yielding high selectivity (*ca* 65%) in the conversion of the cyclohexane to adipic acid (see Figures 4.33[77] and 4.34, which are to be contrasted with Figure 4.31).

From the academic, pedagogic standpoint this example is instructive. The XAFS data are identical for the tetrahedrally coordinated Fe[III] active site in both Fe AlPOs — see Figure 3 of Ref. 76 — but the pore structure is not, and the differences in pore diameter turn out to be crucial in governing the shape selectivity that operates here.

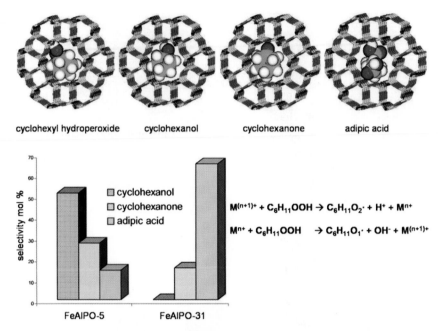

Figure 4.33 | Reaction intermediates in the oxidation of cyclohexane with Fe AlPO-31 (top). The narrower pores (5.4 Å diameter) of Fe AlPO-31 favour the production of adipic acid, compared with the situation that prevails with Fe AlPO-5, and, as a result, adipic acid diffuses out more readily from the pores of Fe AlPO-31 than the other bulkier products. With Fe AlPO-5, the bulkier products diffuse out readily, thus favouring a higher ratio of cyclohexanol and cyclohexanone.[72]

4.7.2.2 *Single-site, molecular-sieve, redox catalysts for the regioselective oxidation of terminal methyl groups in linear alkanes*[75,78]

Saturated hydrocarbons are among the most abundant of all naturally occurring organic molecules, and although they are readily oxidized to completion (burnt) at elevated temperatures, they are also among the most difficult to oxyfunctionalize in a controllable manner at lower temperatures. Their very name, paraffins, from the Latin *parum affinis* (slight affinity), betrays their inertness. Linear alkanes, such as *n*-hexane, resist attack by boiling nitric acid, concentrated sulfuric acid, chromic acid or potassium permanganate. If suitable catalysts were available, it would be environmentally

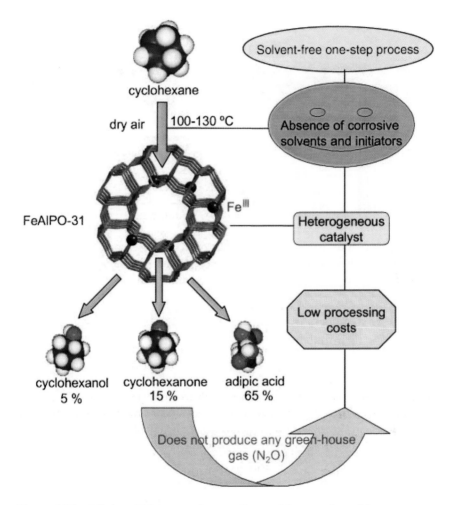

cyclohexane

dry air   100-130 °C

FeAlPO-31

Fe^{III}

Solvent-free one-step process

Absence of corrosive solvents and initiators

Heterogeneous catalyst

Low processing costs

cyclohexanol   cyclohexanone   adipic acid
5 %   15 %   65 %

Does not produce any green-house gas (N$_2$O)

Figure 4.34 Adipic acid is currently manufactured in a costly multi-step process, involving concentrated nitric acid (see Figure 4.31). An alternative one-step, solvent-free process (outlined here) using air and an inexpensive catalyst is shown. (After Thomas *et al.*, *Chem. Eur. J.*, **7**, 2973 (2001).)

acceptable if H$_2$O$_2$ or, better, air and O$_2$ could be used as the oxidants for their oxyfunctionalization. We show in this section that SSHCs of a microporous kind, containing tetrahedrally coordinated Co^{III} or Mn^{III} or Fe^{III} ions, are indeed effective in a regioselective manner to effect "terminal" oxidations of *n*-alkanes.

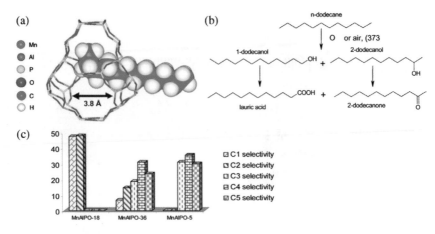

Figure 4.35   Redox active centres in AlPO molecular-sieve catalysts effect regioselective oxyfunctionalization of alkanes at terminal methyl groups. MAPO-18 (M = Co^{III}, Mn^{III}) is especially effective for this purpose.[75]

The first success in our quest to achieve the selective oxidation in O$_2$ of a terminal methyl group in a linear alkane came in our study of *n*-dodecane (see Figure 4.35) where dodecanol, lauric acid and a few other products were formed.[80] Later *n*-pentane, *n*-hexane and *n*-octane were also preferentially converted to the terminal -ol, -al and acids of the corresponding linear alkanes.[79] We selected AlPO-18, which has pores similar to the zeotype analogue of the mineral chabazite. In its framework, a few percent of the redox ions (Co or Mn) may be readily accommodated. We argued[78–80] that oxyfunctionalization would be favoured at the terminal methyl of *n*-alkanes using Co^{III} or Mn^{III} AlPO-18 catalysts because only an end-on approach of the alkane to the active site would be allowed (see Figure 4.36).

With Co AlPO-36 or Mn AlPO-36, because of the much larger pore diameter, there is no such end-on restriction of entry of the alkane into the microenvironment of the active site, which is swathed in molecular O$_2$. In line with the expectations encapsulated in Figure 4.36, where we see that shape selectivity of the *n*-alkane is crucial in the case of M^{III} AlPO-18 but not in M^{III} AlPO-36 or M^{III} AlPO-5, we observe (Figure 4.37) that there is a high preference for terminal (i.e. C$_1$) selectivity in the case of the former catalyst and

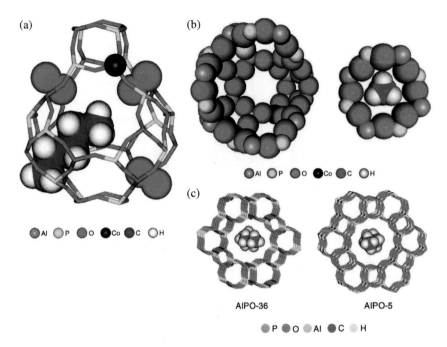

Figure 4.36 (a) Skeletal outline of a single chabazite cage through which molecules of O_2 (red lobes) permeate freely. Alkanes, on the other hand, can enter the chabazite cage by an end-on approach (see bottom left). (b) Views of (left) the interior of the chabazite cage in Co AlPO-18 that is lined with Co^{III} ions, substitutionally replacing framework Al^{III} ions; and (right) the terminal methyl group of a linear alkane (*n*-hexane) which fits snugly into the aperture, where the extremities of the van der Waals radii of the methyl group very nearly touch those of the oxygen atoms of the framework. (c) Representation of the *n*-hexane molecule inside the large-pore AlPO-36 and AlPO-5 structures.[74]

no such selectivity with the latter two. Results for only *n*-octane and *n*-dodecane are shown, but exactly analogous ones, exhibiting high terminal selectivities, are seen for C_6 and C_{10} *n*-alkanes also.

It is well known[73,81] that the aerial oxidation of alkanes by transition metal ions involves the participation of free radicals. We believe that the regioselective oxidation of *n*-alkanes described here (like the shape-selective oxidation of cyclohexane, described in Section 4.7.2.1) is no exception. Because of the larger aperture size of MAPO-36 compared with MAPO-18, it is easier to test the role of free radicals using the former rather than the latter. MAPO-36 has apertures large enough

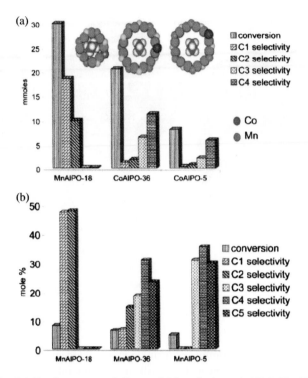

Figure 4.37   (a) Performance of Co- and Mn-substituted AlPO-18, AlPO-36 and AlPO-5 catalysts in the aerial oxidation of *n*-octane after 24 h at 373 K. Because of the end-on approach of the *n*-octane in AlPO-18 only the C_1 and C_2 positions are preferentially oxidized, whereas the larger pore dimensions of AlPO-36 and AlPO-5 lead to oxyfunctionalization predominately at the C_3 and C_4 positions. (b) Regioselective oxyfunctionalization of *n*-dodecane after 24 h at 373 K using Mn AlPO catalysts. The regioselectivity for terminal oxidation with Mn AlPO-18 far exceeds that of the other two catalysts.[75]

to allow alkylperoxyl radicals and radical scavengers to enter their interior surfaces. We have three reasons for believing that the oxyfunctionalization process described here is a free radical process[79]:

- First, there is an induction period in the plot of product yield versus time. It is *ca* 3 h for the reaction at 373 K.
- Second, this induction period is greatly diminished by the prior addition of traces of free radical initiators such as *t*-butyl

hydroperoxide (TBHP). Moreover addition of small quantities of hydroquinone, a free radical scavenger, greatly hinders the oxidation and prolongs the induction period.

- Third, in the oxidation of cyclohexane (the peroxide of which is easier to analyse than that of *n*-alkanes) under identical conditions over a range of Co^{III} and Mn^{III} AlPO catalysts, the presence of cyclohexyl hydroperoxide (which subsequently decomposes to cyclohexanol and cyclohexanone) has been directly detected (see Figure 4.33).

By boosting the degree of isomorphous replacement of framework Al^{III} ions by either Co^{III} or Mn^{III} from 4 to 10 at %, so that two redox-active sites are placed opposite one another in the same cage of the chabazite (MAPO-18) solid, it is possible[76] to oxidize selectively each terminal CH_3 group of the *n*-hexane in sequence, thereby producing adipic acid (as illustrated in Figure 4.38). This is another example of cascade (consecutive) catalysis.

It is relevant here to recall the so-called Barton challenge, issued by American organic chemist J. D. Roberts in 1999. During D. H. R. Barton's last ten years as an organic chemist, he was preoccupied with working out experimental ways of oxidizing a C-H bond in a CH_3 group (the dissociation energy of which is 104 kcal mol^{-1}) in preference to oxidizing a C-H bond in either a CH_2 group (dissociation energy 95 kcal mol^{-1}) or in a tertiary C-H (dissociation energy 91 kcal mol^{-1}). The goal requested by Roberts is 80% conversion to adipic acid. Our own work, illustrated in the lower half of Figure 4.38, has reached *ca* 60%.

4.7.2.3 *Other important selective oxidations with redox SSH microporous catalysts*

The process of C-H activation such as that which we have described in cyclic and linear alkanes may be extended to comparable situations in related molecules. These include:

- The aerial conversion[82] of toluene to benzaldehyde and benzoic acid.

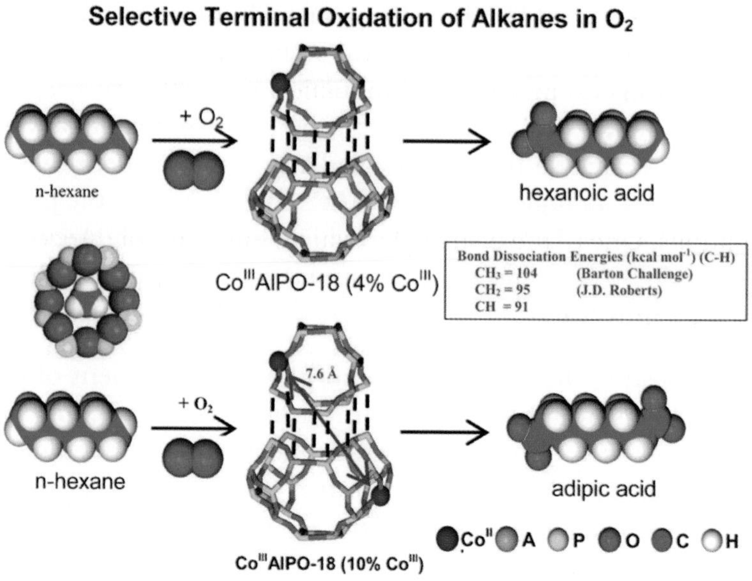

Figure 4.38 Depending upon the concentration and location of the redox active centres in Co^{III} AlPO-18 molecular-sieve catalysts, a linear alkane (like hexane) may be selectively oxidized to either *n*-alkanoic acid (top) or to the di- (terminal) carboxylic acid (bottom).[75]

- The preparation of vitamin B_3 (niacin)[83] from 3-picoline, as well as several other pharmaceutically important products.[84]

And they are both of great commercial importance. We discuss these conversions, carried out via SSHCs of a microporous kind, as well as other industrially significant ones in Chapter 5.

4.8 Insights from Quantum Chemical Computations into the Mechanism of C-H Activation at Mn^{III} Catalytic Centres in Microporous Solids

Apart from demonstrating the free radical nature of the aerial oxidation of alkanes at Mn^{III}, Co^{III} and Fe^{III} active sites in microporous catalysts doped with (framework) transition metal ions, experimentalists have, to date, failed to identify all the short-lived, unstable intermediates that are involved in the conversions described in Section 4.7. Notwithstanding some helpful kinetic studies by Iglesia and co-workers,[85] and the indisputable fact of the occurrence of induction periods prior to the conversion of the alkanes to the corresponding alkanols, alkanals and acids, no one could hitherto formulate a satisfactory detailed mechanism for the selective aerobic oxidations. One of the principal obstacles was a lack of direct experimental means to interrogate, with *in situ* methods, the oxidation at such high pressures (20 bar typically), especially to identify the putative participation of many short-lived intermediates, both bound and free.

Very recently, four seminal papers have appeared that shed great light on the initial and subsequent stages of aerobic selective oxidation of hydrocarbons by Mn^{III}-doped AlPOs.[86,87] By applying state-of-the-art electronic structure techniques, based on hybrid exchange functionals in DFT and periodic boundary conditions, Gomez-Hortiguela *et al.*[86,87] have unravelled the reaction mechanism responsible for the individual and overall processes involved.

The quantum chemical approach entailed choosing AlPO-5 (AFI) as the archetype of the Mn^{III}-doped SSHCs. This structure comprises a one-dimensional non-interconnected 12-membered ring channel with only one type of tetrahedral site into which the Mn ion is accommodated. The Mn-doped framework was described with periodic boundary conditions using the P1 space group, which has 72 atoms in the crystallographic unit cell. Gomez-Hortiguela *et al.*[86] "inserted" one Mn ion (to replace an Al ion) per unit cell, and performed DFT calculations as implemented in the CRYSTAL program of Dovesi *et al.* Their results clearly identify a so-called preactivation step as well as a propagation cycle. Both of these and all the intermediate reactants and products involved are shown in Figure 4.39. Enthalpies and activation energies (shown in red and black respectively in this figure) were calculated for all steps.

In the preactivation step of the catalyst, which precedes the catalytic propagation cycle in which the final products of oxidation (alcohol, aldehyde and acid) are formed, one of the Mn^{III} ions initially present in the catalyst is transformed by interaction with an

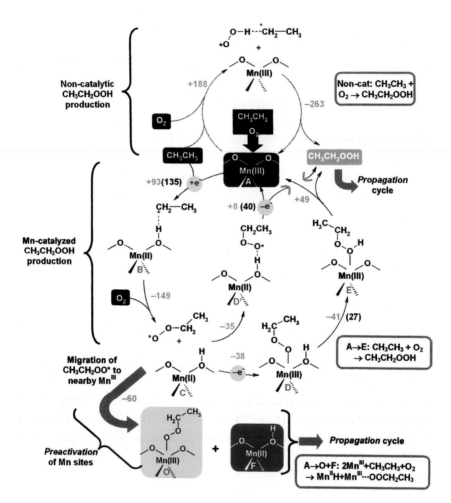

Figure 4.39 Preactivation mechanism (A–E) for the production of CH_3CH_2OOH from RH and O_2 without (top) and with the assistance of Mn^{III} (middle). A black background indicates initial catalyst and reactant molecules, and a red background highlights the hydroperoxide intermediate. Other background colours (yellow, blue) are used to indicate intermediates produced here that are necessary to initiate the subsequent propagation cycle. This cycle also yields the activation of Mn sites through reduction of Mn^{III} to Mn^{II} (F, with blue background) and formation of the ROO...Mn^{III} complex (O, with yellow background) (A–O + F) after migration of ROO radicals. Enthalpies (red) and activation energies, if any (black, in brackets), are shown for each supplementary step, in kilojoules per mole.[86]

alkane — ethane was chosen for simplicity — and one O_2 molecule into new Mn-bearing species: a reduced (Mn^{II}) site as well as a Mn^{III}...peroxo complex, which are active for the subsequent propagation cycle. It is interesting that the preactivation step has a high activation energy, calculated as 135 kJ mol^{-1} (see step A to B in Figure 4.39), which is in line with the long induction period observed experimentally by Thomas *et al.*[79]

The calculations of Gomez-Hortiguela *et al.*[86,87] also reveal that Mn^{III} sites are able to produce the hydroperoxide intermediate from C_2H_6 and O_2. However, this intermediate can be transformed into the oxidized products only through the agency of Mn^{II} sites, which are formed from Mn^{III} sites in the preactivation step via a H-abstraction from the ethane, thereby also yielding an alkyl ($CH_3CH_2\bullet$) radical. This radical subsequently adds O_2 in a stereospecific manner to form a free peroxoradical $CH_2CH_5OO\bullet$ (step B to C in Figure 4.39). Migration of the $C_2H_5OO\bullet$ radical in the nanopores of AlPO-5 frees Mn^{III} for the propagation cycle and forms Mn^{II}...$C_2H_5OO\bullet$ complexes, which are also needed for propagation.

In summary, the Gomez-Hortiguela papers demonstrate that the catalytic efficiency of Mn^{III} AlPO single-site catalysts in selective aerobic oxidation is intrinsically linked to (i) the Mn redox activity between valence states III and II, and (ii) the coordinative unsaturation of tetrahedral Mn ions substituted into the AlPO frameworks, which facilitate the reactions by stabilizing oxo-type radicals through the formation of complexes with the Mn ions. Their derived mechanism accounts well for the crucial role of both Mn^{III} and Mn^{II} in the overall reaction. It also shows that hydrocarbon activation catalysed by Mn ions requires much lower activation energies than through non-catalytic pathways — see top of Figure 4.39, where it is seen that the enthalpy change required for such a pathway is as high as 188 kJ mol^{-1}.

It is also noteworthy that the Gomez-Hortiguela *et al.*[87] results explicitly rule out a mechanism involving active participation of a Mn^{III}...O_2 complex. Their geometry optimizations always resulted in O_2 moving away from the Mn, regardless of the O_2 spin and orientation, with Mn returning to its tetrahedral environment and the resistance of the Mn^{III} elevated to a higher oxidation state such as Mn^{IV}. This, too, is in line with the experimental work reported in Ref. 79.

The convincing work of Gomez-Hortiguela *et al.*[86,87] has yielded detailed knowledge of the reaction mechanism of the aerobic selective oxidation of hydrocarbons catalysed by four-coordinate Mn^{III} single sites in an AlPO matrix. It confirms and complements the data available from experiments. It also draws attention to the fact that, with such complex reactions, experiments alone cannot provide the full mechanistic picture, owing to the short lifetime, instability and complexity of the implicated intermediates.

The stage is now set for a similar investigation of the known catalytic performance of Co^{III} and Fe^{III} substituted into AlPOs as single-site catalysts. Is it possible that, for such catalysts, higher oxidation states (e.g. Fe^{IV} or Fe^{V}) may be involved, as is known in P450 chemistry. One also wishes to know if, in a mixed-dopant system, where, for example, both Mn^{III} and Fe^{III} are present and spatially well separated in tetrahedral coordination, all the Fe would be oxidized (and into which oxidation state?) and all the Mn would be in the Mn^{II} state. It would also be instructive if quantum chemical computations such as those described in Refs 86 and 87 were applied to the way in which an Mn^{III} or Co^{III} catalytic centre (in AlPOs) readily produces (*in situ*) hydroxylamine (NH_2OH) from NH_3 and O_2 at relatively low temperatures. This is a key reaction in a new method for producing the oxime of cyclohexanone, a vital stepping stone in the synthesis of nylon 6, as we discuss fully in Chapter 5.

Finally, Gomez-Hortiguela *et al.* ended their investigation by asserting that "such a complete understanding of the catalytic mechanism is fundamental to design more efficient oxidation catalysts". Whilst this statement contains much truth, it must not be forgotten that the Mn^{III}-based catalyst was itself designed[80] by an entirely experimental (solid-state chemical) approach reminiscent of the discovery of other SSHCs that are described in earlier sections of this chapter.

4.9 Bifunctional Single-site Microporous Catalysts: A Solvent-free Synthesis of Caprolactam, the Precursor of Nylon 6

Many heterogeneous reactions utilized in the most important industrial processes (like reforming, cracking and general hydroprocessing of hydrocarbons) require the presence of more than one type of catalyst to achieve a significant yield of desired product. The conversion of *n*-heptane to isoheptane, for example, requires the presence of a dehydrogenation catalyst (such as Pt supported on Al_2O_3, or a Pt-Ir bimetallic catalyst) supported on an acidic solid such as a porous silica-alumina or zeolite. The *n*-heptane is dehydrogenated by the Pt or Pt-Ir to *n*-heptene which, in turn, is isomerized to isoheptene by the SiO_2-Al_2O_3 or zeolite catalyst. The final step is the hydrogenation of isoheptene to isoheptane in the presence of the Pt. In principle, each of these reaction stages could be accomplished separately; alternatively, they could be effected within the same environment by mixing together the appropriate amounts of each catalyst, or even by dispersing the platinum group metal(s) on the porous SiO_2-Al_2O_3. Porous catalysts are normally employed on a massive industrial scale[88] to achieve this dual function because the large surface areas available enhance the throughput of product.

The supreme advantage that microporous solids of the type discussed in this chapter possess is that, for certain important reactions, the two distinct catalytic sites can be incorporated into one solid: we have one crystallographic phase not two, as in the cases outlined in the preceding paragraph. An appropriately designed nanoporous solid containing spatially well-separated active sites of two distinct kinds is the desired goal.

This goal has been achieved by Thomas and Raja[4,89] in their solvent-free, environmentally benign method of synthesizing ε-caprolactam, the precursor of (the recyclable — see Chapter 5) nylon 6. Currently, one of the popular industrial methods of producing caprolactam entails a two-step process, each of which uses expensive, corrosive and environmentally harmful reagents — see Scheme 4.7.

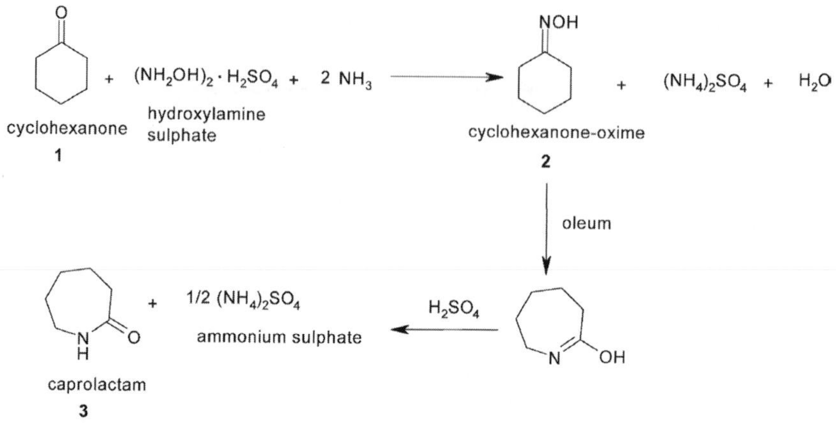

Scheme 4.7 A current, popular method for the manufacture of caprolactam, using aggressive reagents.[4]

Our aim when we embarked[89,90] on the task of devising a low-temperature, solvent-free, benign method of synthesizing ε-caprolactam was to identify a family of SSHCs that could be controllably tuned to yield maximum efficiencies in regard to the production of (i) hydroxylamine, NH_2OH, under *in situ* conditions and (ii) ε-caprolactam. The desiderata for a reliable and effective bifunctional catalyst are as follows:

- isolated redox active sites which can produce NH_2OH *in situ*, from NH_3 and air;
- isolated Brønsted acid active sites which will isomerize (by Beckmann rearrangement) the ammoximation product (oxime) formed from cyclohexanone and NH_2OH;
- a sufficiently open structure so as to facilitate the diffusion of all the reactant molecules (to gain access to the active sites) and products within the catalyst;
- a reproducible, efficient preparative method for obtaining a phase-pure, open structure which retains its (two kinds) of active sites, without leaching out during use; and
- a method of structural characterization which can determine the effectiveness of the single-site nature of the bifunctional catalyst.

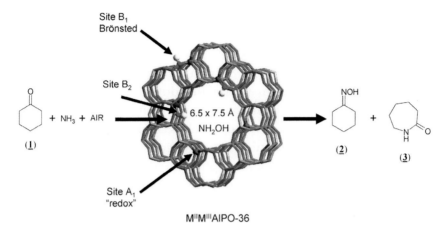

Figure 4.40 At the oxidizing active centre, Site A_1, ammonia is converted to NH_2OH, which converts the cyclohexanone (1) to the oxime (2), which is then catalytically converted to ε-caprolactam (3) at the acidic active site (adjacent to the Mg^{II}). (Colour code: green, Co^{III}; blue, Mg^{II}; white, H.)[4]

All these desiderata could be met[89] by taking strategically designed microporous AlPOs as receptacles, and by incorporating, during synthesis, in the presence of structure-directing agents, the active sites required for the sequence of steps outlined in Figure 4.40.

We shall return to caprolactam and nylon 6 production, and related considerations, more fully in Chapter 5. Suffice it to say that one of the best combinations for the $M^{II}M^{III}$ AlPO-36 (see Figure 4.40) is that which has Mg^{II} as M^{II} and Mn^{III} as M^{III}.

4.10 Single-site Metal Cluster Catalysts Supported on a Microporous Host: Reactive Environments Influence the Structure of Catalysts

Here we focus mainly on the work of Gates and co-workers[91–93] who have made detailed studies of Ir_4 and Ir_6 clusters on dealuminated zeolite Y (Si/Al *ca* 30). Before proceeding to summarize their results, which are especially relevant to the author's own work on silica-supported bimetallic clusters and are described in Chapter 8, we first recall some key facts pertaining to the phenomenon of catalyst structural modification wrought by the presence of the reactant species.

Earlier in this chapter (see Section 4.2) we saw evidence for the migration of Ni ions from within the double-six rings (in extraframework positions) of faujasitic zeolite out into the supercage under the influence of acetylene, which is catalytically trimerized by the Ni ions.[12,13] This structural modification during the course of catalytic turnover is a well-known phenomenon and has been studied at a fundamental level — on catalysts quite distinct from those that constitute the subject of this monograph — by Ertl and co-workers,[94,95] by workers at the Topsøe Research Center[96] and by Gai and colleagues.[97–99]

In the detailed studies of Gates and co-workers,[91] it was conclusively established that the Ir_4 clusters, represented ideally as in Figure 4.41, break up during the process of ethene hydrogenation, when the catalysis is conducted with an equimolar ratio of C_2H_4/H_2 into mononuclear Ir entities. But when the ratio of ethene to H_2 is decreased to 0.3 (i.e. H_2-rich), the tetranuclear clusters reassemble. In other words, by controlling the composition of the reactant gases (at a temperature of 353 K) one may switch the nuclearity of the Ir catalyst between 4 and unity. The *in situ* technique used in this work

Figure shows a representative Ir_4 cluster supported on an $Al_3Si_3O_6H_{15}$ model representing a zeolite X- supercage.

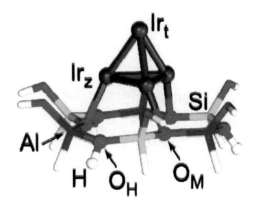

Figure 4.41 Schematic representation of the siting of the Ir_4 cluster anchored to dealuminated zeolite Y.[92]

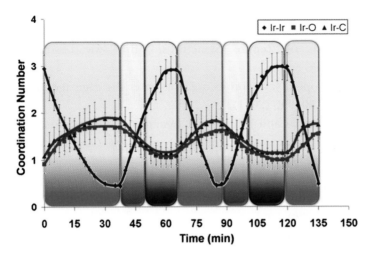

Figure 4.42 The EXAFS results of Uzun and Gates[91] in their *in situ* study of the changes in Ir-Ir, Ir-O and Ir-C coordination numbers. The first (red) rectangle shows how the Ir_4 clusters become Ir species during exposure to a $C_2H_4/H_2 = 4$ ratio of gas mixture. When this ratio is changed to 0.3, after 35 min the Ir_4 clusters reform. After 60 min, they break down again on reverting to the ratio of 4. (After Uzun *et al.*[93])

is time-resolved EXAF5,[100,101] which readily yields the Ir-Ir coordination number from 3 (for Ir_4) to an immeasurably small value, for individual Ir atoms bound to the microporous support.

When the feed composition was cycled from ethene-rich to H_2-rich, the predominant species in the catalyst cycled accordingly, as shown in Figure 4.42. This is the first example of a study in which the structure of the supported nanocluster catalyst may be controllably adjusted by modifying the composition of the reacting gas.

It will be seen in Chapter 8 that, following on from the pioneering work of Sinfelt with nanoparticle bimetallic catalysts (where there may be as many as 10^7 or more atoms in the nanoparticles), nanocluster bimetallic catalysts, in which there may be no more (and often fewer) than about 20 atoms in all — so that they behave, as do Gates' Ir_4 and Ir_6 species, rather more like molecules than metals — are exceptionally active single-site multinuclear catalysts.

References

1. Database of Zeolite Structures, http://www.iza-structure.org/databases/.
2. A. I. Cooper. Nanoporous organics enter the cage age, *Angew. Chem. Int. Ed.*, **50**, 996 (2011).
3. J. M. Thomas and J. Klinowski. Systematic enumeration of microporous solids: towards designer catalysts, *Angew. Chem. Int. Ed.*, **46**, 7160 (2007).
4. R. Raja and J. M. Thomas. Nanoporous solids as receptacles and catalysts for unusual conversions of organic compounds, *Solid State Sci.*, **8**, 326 (2006).
5. J. M. Thomas. Uniform heterogeneous catalysts: the role of solid-state chemistry in their development and design, *Angew. Chem. Int. Ed.*, **27**, 1673 (1988).
6. Ch. Baerlocher, L. B. McCusker and D. H. Olson. *Atlas of Zeolite Framework Types*, 6th revised ed., Elsevier, Amsterdam (2007).
7. J. M. Thomas, R. G. Bell and C. R. A. Catlow. A synoptic guide to the structures of zeolitic and related solid catalysts, in *Handbook of Heterogeneous Catalysis*, ed. G. Ertl, H. Knözinger and J. Weit-Kamp, Vol. 1, Wiley-VCH, Weinheim (1997), p. 286.
8. J. M. Thomas and J. Klinowski. The study of aluminosilica and related catalysts by high-resolution solid-state NMR, *Adv. Catal.*, **33**, 199 (1985).
9. F. Eder, M. Stockenhuber and J. A. Lercher. Sorption of light alkanes on H-ZSM-5 and H-mordenite, in *Zeolites: A Refined Tool for Designing Catalystic Sites*, ed. L. Bonneviot and S. Kaliaguine, Elsevier, Amsterdam, (1995).
10. A. Bhan, A. D. Allian, G. J. Sunley, D. J. Law and E. Iglesia. Specificity of sites within eight-membered ring zeolite channels for carbonylation of methyls to acetyls, *J. Am. Chem. Soc.*, **129**, 4919 (2007).
11. J. W. Couves, J. M. Thomas, D. Waller, R. H. Jones, A. J. Dent, G. E. Derbyshire and G. N. Greaves. Tracing the conversion of aurichalcite to a copper catalyst by combined X-ray adsorption and diffraction, *Nature*, **354**, 465 (1991).
12. (a) P. J. Maddox, J. Stachurski and J. M. Thomas. Probing structural changes during the onset of catalytic activity by *in situ* X-ray diffractometry, *Catal. Lett.*, **1**, 191 (1988).

(b) J. M. Thomas, C. Williams and T. Rayment. Monitoring cation-site occupancy of nickel-exchanged zeolite Y catalysts by high-temperature *in situ* X-ray powder diffractometry, *J. Chem. Soc. Faraday Trans.*, **84**, 2915 (1988).

13. A. R. George, C. R. A. Catlow and J. M. Thomas. Catalytic cyclotrimerization of acetylene: a computational study, *J. Chem. Soc. Faraday Trans.*, **91**, 3975 (1995).

14. P. A. Wright, private communication, November 2010. See also *Microporous Framework Solids*, RSC Publishing, London (2009).

15. I. Bull, G. S. Koermer, A. Moini and S. Unverricht. Catalysts, systems and methods utilizing non-zeolitic metal-containing molecular sieves having the CHA crystal structure, U.S. Patent No. 2009/0196812A1 (2009).

16. (a) J. M. Thomas. Design, synthesis and *in situ* characterization of new solid catalysts, *Angew. Chem. Int. Ed.*, **38**, 3588 (1999).

    (b) J. M. Thomas. Solid acid catalysts, *Sci. Am.*, **266**, 85 (1992).

17. A. K. Cheetham, M. M. Eddy and J. M. Thomas. The direct observation of cation hydrolysis in lanthanum zeolite-Y by neutron diffraction, *Chem. Commun.*, 1337 (1984).

18. M. Boronat and A. Corma. Are carbenium and carbonium ions reaction intermediates in zeolite-catalyzed reactions?, *Appl. Catal. A. Gen.*, **336**, 2 (2008).

19. K. Smith, A. Musson and G. A. DeBoos. A novel method for the nitration of simple aromatic compounds, *J. Org. Chem.*, **63**, 8448 (1998).

20. K. Smith, Z. Zhao and P. K. G. Hodgson. Synthesis of aromatic ketones by acylation of aryl ethers with carboxylic anhydrides in the presence of zeolite H-β (H-BEA) in the absence of solvent, *J. Mol. Cat. A. Chem.*, **134**, 121 (1998).

21. K. Smith and S. D. Roberts. Regioselective dialkylation of naphthalene, *Catal. Today*, **60**, 227 (2000).

22. R. A. Sheldon, I. Arends and U. Hanefeld. *Green Chemistry and Catalysis*, Wiley-VCH, Weinheim (2007).

23. J. M. Thomas, J. C. Hernandez-Garrido and R. G. Bell. A general strategy for the design of new solid catalysts for environmentally benign conversions, *Top. Catal.*, **52**, 1630 (2009).

24. D. E. W. Vaughan, K. G. Strohmaier, W. J. Murphy, I. A. Cody and S. J. Linek. Silicoaluminophosphates having an AEI structure: a

method for their preparation and their use as catalysts for the hydro-processing of hydrocarbons, U.S. Patent No. 6238550 BI (2001).

25. R. A. van Santen and M. Neurock. *Molecular Heterogeneous Catalysis: A Conceptual and Computational Approach*, Wiley-VCH, Weinheim (2006).

26. B. Smit and T. L. M. Maesen. Towards a molecular understanding of shape selectivity, *Nature*, **457**, 671 (2008).

27. J. I. Siepmann and D. Frenkel. Configurational-bias Monte Carlo: a new sampling scheme for flexible chains, *Mol. Phys.*, **75**, 59 (1998).

28. S. Yashonath, J. M. Thomas, A. K. Nowak and A. K. Cheetham. The siting, energetics and mobility of saturated hydrocarbons inside zeolitic cages: methane in zeolite-Y, *Nature*, **331**, 601 (1988).

29. S. D. Pickett, A. K. Nowak, J. M. Thomas, B. K. Peterson, J. F. P. Swift, A. K. Cheetham, C. J. J. den Ouden, B. Smit and M. M. F. Post. Mobility of adsorbed species in zeolites: a molecular dynamics simulation of xenon in silicalite, *J. Phys. Chem.*, **94**, 1233 (1990).

30. A. J. Paterson. Ph.D. Thesis, University of Southampton, January 2011.

31. M. Lefenfeld, R. Raja, A. J. Paterson and M. E. Polter. Catalytic dehydration of alcohols using phase-pure calcined single- and multi-site heterogeneous catalysts, WO/2010085708 (2010).

32. Y. Xu, C. P. Grey, J. M. Thomas and A. K. Cheetham. An investigation into the conversion of methanol to hydrocarbons over SAPO-34 catalyst using MASNMR and gas chromatography, *Catal. Lett.*, **4**, 251 (1990).

33. U. Olsbye, S. Svelle, M. Bjørgen, P. Beato, S. Bordiga and K. P. Lillerud. Methanol to gasoline or olefins — how zeolite dimensionality and pore size controls product selectivity, *Angew. Chem. Int. Ed.* (unpublished work) (2011).

34. J. Chen and J. M. Thomas. MAPO-18 (M = Mg, Zn, Co): a new family of catalysts for the conversion of methanol to light olefins, *Chem. Commun.*, 603 (1994).

35. N. Y. Chen and W. J. Reagen. Evidence of autocatalysis in methanol to hydrocarbons reaction over zeolite catalysts, *J. Catal.*, **59**, 123 (1979).

36. I. M. Dahl and S. Kolboe. On the reaction mechanism for hydrocarbon formation from methanol over SAPO-34. 1. Isotopic labeling studies of the co-reaction of ethane and methanol, *J. Catal.*, **149**, 458 (1994).

37. D. M. McCann, D. Lesthaeghe, P. W. Kletnieks, D. R. Guenther, M. J. Hayman, V. van Speybroeck, M. Waroquier and J. F. Haw. A complete catalytic cycle for supramolecular methonal-to-olefin conversion, *Angew. Chem. Int. Ed.*, **47**, 5179 (2008).

38. S. Svelle, F. Joensen, J. Nerlov, U. Olsbye, K. P. Lillerud, S. Kolboe and M. Bjørgen. Conversion of methanol into hydrocarbons over zeolite H-ZSM-5: ethene formation is mechanistically separated from the formation of higher alkenes, *J. Am. Chem. Soc.*, **128**, 14770 (2006).

39. S. Teketel, S. Svelle, K. P. Lillerud and U. Olsbye. Shape-selective conversion of methanol to hydrocarbons over 10-ring unidirectional-channel acidic H-ZSM-22, *Chem. Cat. Chem.*, **1**, 78 (2009).

40. W. Dai, W. Kong, L. Li, G. Wu, N. Guan and N. Li. The effect of organic impurities originating from the incomplete combustion of organic templates on the methanol-to-olefins reaction over SAPO-46, *Chem. Cat. Chem.*, **2**, 1548 (2010).

41. W. Dai, X. Wang, G. Wu, N. Guan, M. Hunger and L. Li. Methanol-to-olefin conversion on silicoaluminophosphate catalysts: effect of Brønsted acid sites and framework structures, *ACS Catal.*, **1**, 4 (2011).

42. K. I. Zamaraev and J. M. Thomas. Structural and mechanistic aspects of the dehydration of isomeric butyl alcohols over porous aluminosilicate acid catalysts, *Adv. Catal.*, **41**, 335 (1996).

43. M. A. Makarova, C. Williams, V. N. Romannikov, K. I. Zamaraev and J. M. Thomas. Influence of pore confinement or the catalytic dehydration of isobutanol on H-ZSM-5, *J. Chem. Soc. Faraday Trans.*, **86**, 581 (1990).

44. M. A. Makarova, C. Williams, J. M. Thomas and K. I. Zamaraev. Dehydration of *n*-butanol on HNa-ZSM-5, *Catal. Lett.*, **4**, 261 (1990).

45. C. Williams, M. A. Makarova, L. V. Malysheva, E. A. Paukshtis, K. I. Zamaraev and J. M. Thomas. Mechanistic studies of the catalyzed dehydration of iso-butanol on Na-H-ZSM-5, *J. Chem. Soc. Faraday Trans.*, **86**, 3473 (1990).

46. C. Williams, M. A. Makarova, L. V. Malysheva, E. A. Paukshtis, E. P. Talsi, J. M. Thomas and K. I. Zamaraev. Kinetic studies of catalytic dehydration of *tert*-butanol on zeolite NaH-ZSM-5, *J. Catal.*, **127**, 377 (1991).

47. M. A. Makarova, E. A. Paukshtis, J. M. Thomas, C. Williams and K. I. Zamaraev. *In Situ* FTIR kinetic studies of diffusion, adsorption and dehydration reactions of *t*-butanol in zeolite NaH-ZSM-5, *Catal. Today*, **9**, 61 (1991).

48. A. G. Stepanov, K. I. Zamaraev and J. M. Thomas. ^{13}C CP/MAS and ^2H NMR study of *t*-butylalcohol dehydration on H-ZSM-5: evidence for the formation of *t*-butyl silyl ether intermediates, *Catal. Lett.*, **13**, 407 (1992).

49. J. Sommer and R. Jost. Carbenium and carbonium ions in liquid- and solid-superacid-catalyzed activation of small alkanes, *Pure Appl. Chem.*, **72**, 2309 (2000).

50. C. Lamberti, E. Groppo, G. Spoto, S. Bordiga and A. Zecchina. Infrared spectroscopy of transient surface species, *Adv. Catal.*, **51**, 54 (2007).

51. See B. Notari. Microporous crystalline titanium silicates, *Adv. Catal.*, **41**, 253 (1996), and references therein.

52. J. M. Thomas and G. Sankar. The role of synchrotron-based studies in the elucidation and design of active sites in titanium-silica epoxidation catalysts, *Acc. Chem. Res.*, **34**, 571 (2001).

53. A. Zecchina, S. Bordiga, G. Spoto, A. Damin, G. Berlier, F. Bonino, C. Prestipino and C. Lamberti. *In situ* characterization of catalysts active in partial oxidatives TS-1 and Fe-MFI case studies, *Top. Catal.*, **21**, 67 (2002).

54. D. Gleeson, G. Sankar, C. R. A. Catlow, J. M. Thomas, G. Spano, S. Bordiga, A. Zecchina and C. Lamberti. The architecture of catalytically active centers in titanosilicate (TS-1) and related selective oxidation catalysts, *Phys. Chem. Chem. Phys.*, **2**, 4812 (2000).

55. P. Ratnasamy, D. Srinivas and H. Knözinger. Active sites and reactive intermediates in titanium silicate molecular sieves, *Adv. Catal.*, **48**, 1 (2004), and references therein.

56. M. A. Camblor, A. Corma, A. Martinez and J. Perez-Pariente. Synthesis of a titaniumsilicoaluminate isomorphous to zeolite Beta and its application as a catalyst for the selective oxidation of large molecules, *Chem. Commun.*, 589 (1992).

57. A. Tuel and Y. Ben Taarit. Synthesis and catalytic properties of titanium-substituted silicoaluminophosphate TAPSO-5, *J. Chem. Soc. Chem. Commun.*, 1667 (1994).

58. N. Ulagappan and V. Krishnasamy. Titanium substitution in silicon-free molecular sieves: anatase-free $TAPO_4$-5 and $TAPO_4$-11 synthesis and characterisation for hydroxylation of phenol, *J. Chem. Soc. Chem. Commun.*, 373 (1995).

59. A. Corma, M. J. Díaz-Cabañas, M. E. Domine and F. Rey. Ultra fast and efficient synthesis of Ti-ITQ-7 and positive catalytic implications, *Chem. Commun.*, 1725 (2007).

60. P. Wu, T. Tatsumi and T. Yashima. A new generation of titanosilicate catalyst for epoxidation of alkenes, *J. Phys. Chem. B*, **105**, 2897 (2001).

61. A. Corma. Attempts to fill the gap between enzymatic, homogeneous and heterogeneous catalysis, *Catal. Rev.*, **46**, 369 (2004).

62. E. Taarning, S. Saravanamurugan, M. S. Holm, J. Xiong, R. M. West and C. H. Christensen. Zeolite-catalyzed isomerization of triose sugars, *Chem. Sus. Chem.*, **2**, 625 (2009).

63. M. S. Holm, S. Saravanamurugan and E. Taarning. Conversion of sugars to lactic acid derivatives using heterogeneous zeotype catalysts, *Science*, **328**, 602 (2010).

64. M. Moliner, Y. Roman-Leshkov and M. E. Davis. Tin-containing zeolites are highly active catalysts for the isomerization of glucose in water, *PNAS*, **107**, 6164 (2010).

65. S. O. Lee, R. Raja, K. D. M. Harris, J. M. Thomas, B. F. G. Johnson and G. Sankar. Mechanistic insights into the conversion of cyclohexene to adipic acid by H_2O_2 in the presence of a TAPO-5 catalyst, *Angew. Chem. Int. Ed.*, **42**, 1520 (2003).

66. J. M. Thomas and R. Raja. Mono-, bi- and multi-functional single-sites: exploring the interface between homogeneous and heterogeneous catalysis, *Top. Catal.*, **53**, 848 (2010).

67. W. F. Hoelderich. 'One-pot' reactions: a contribution to environmental protection, *Appl. Catal. A*, **194**, 487 (2000).

68. G. I. Panov, V. I. Sobolev and A. Skhanitonov. The role of iron in N_2O decomposition on ZSM-5 zeolite and reactivity of the surface oxygen formed, *J. Mol. Catal.*, **61**, 85 (1980).

69. L. Marchese, J. Chen, J. M. Thomas, S. Coluccia and A. Zecchina. Brønsted, Lewis and redox centers in CoAlPO-18 catalysts: vibrational modes of adsorbed water, *J. Phys. Chem.*, **98**, 13350 (1994).

70. P. B. Barrett, G. Sankar, C. R. A. Catlow and J. M. Thomas. X-ray absorption spectroscopic study of Brønsted, Lewis and redox centers in cobalt-substituted aluminium phosphate catalysts, *J. Phys. Chem.*, **100**, 8977 (1996).

71. J. M. Thomas and G. Sankar. *In situ* combined X-ray absorption spectroscopy and X-ray diffractometric studies of solid catalysts, *Top. Catal.*, **8**, 1 (1999).

72. V. L. Pecoraro (ed.). *Manganese Redox Enzymes*, VCH, Weinheim (1992).

73. C. L. Hill (ed.). *Activation and Functionalization of Alkanes*, Wiley, Chichester (1989).

74. P. A. MacFaul, K. U. Ingold, D. D. M. Wayner and L. Que Jr. A putative monooxygenase mimic which functions via well-disguised free radical chemistry, *J. Am. Chem. Soc.*, **119**, 10594 (1997).

75. J. M. Thomas, R. Raja, G. Sankar and R. G. Bell. Molecular sieve catalysts for the regioselective and shape-selective oxyfunctionalization of alkanes in air, *Acc. Chem. Res.*, **34**, 191 (2001).

76. R. Raja, G. Sankar and J. M. Thomas. Powerful redox molecular sieve catalysts for the selective oxidation of cyclohexane in air, *J. Am. Chem. Soc.*, **121**, 11926 (1999).

77. M. Dugal, G. Sankar, R. Raja and J. M. Thomas. Designing a heterogeneous catalyst for the production of adipic acid by aerial oxidation of cyclohexane, *Angew. Chem. Int. Ed.*, **39**, 2310 (2000).

78. J. M. Thomas and R. Raja. The advantages and future potential of single-site heterogeneous catalysts, *Top. Catal.*, **40**, 3 (2006).

79. J. M. Thomas, R. Raja, G. Sankar and R. G. Bell. Molecular-sieve catalysts for the selective oxidation of linear alkanes by molecular oxygen, *Nature*, **398**, 227 (1999).

80. R. Raja and J. M. Thomas. A manganese-containing molecular sieve catalyst designed for the terminal oxidation of dodecane in air, *Chem. Commun.*, 1841 (1998).

81. T. Maschmeyer, J. M. Thomas, G. Sankar, R. D. Oldroyd, I. J. Shannon, J. A. Klepto, A. F. Masters, J. K. Beattie and C. R. A. Catlow. Designing a solid catalyst for the selective low-temperature oxidation of cyclohexane to cyclohexanone, *Angew. Chem. Int. Ed.*, **36**, 1639 (1997).

82. R. Raja, J. M. Thomas and V. R. Dreyer. Benign oxidants and single-site solid catalysts for the solvent-free selective oxidation of toluene, *Catal. Lett.*, **110**, 179 (2006).

83. R. Raja, J. M. Thomas, M. Greenhill-Hooper, S. V. Ley and F. A. A. Paz. Facile, one-step production of niacin (vitamin B_3) and other nitrogen-containing phamarceuticals with a single-site solid catalyst, *Chem. Eur. J.*, **14**, 2340 (2008).

84. J. M. Thomas and R. Raja. Designed, open-structure heterogeneous catalysts for the synthesis of fine chemicals and pharmaceuticals, *Stud. Surf. Sci. Catal.*, **170a**, 19 (2007).

85. B. Moden, B. Z. Zhou, J. Dakka, J. G. Santiesteban and E. Iglesia. Kinetics and mechanism of cyclohexane oxidation on Mn AlPO-5, *J. Catal.*, **239**, 390 (2006).

86. L. Gomez-Hortiguela, F. Cora, G. Sankar, C. M. Zicovich-Wilson and C. R. A. Catlow. Catalytic reaction mechanism of Mn-doped nanoporous aluminophosphate for the aerobic oxidation of hydrocarbons, *Chem. Eur. J.*, **16**, 1368 (2010).

87. L. Gomez-Hortiguela, F. Cora and C. R. A. Catlow. Aerobic oxidation of hydrocarbons catalyzed by Mn-doped nanoporous aluminophosphates: preactivation of the Mn sites, *ACS Catal.*, **1**, 18 (2011); see also *ACS Catal.*, **1**, 945 (2011).

88. J. M. Thomas and W. J. Thomas. *Principles and Practice of Heterogeneous Catalysis*, Wiley-VCH, Weinheim (1997) — see, in particular Section 6.1.2 (pp. 427 *et seq.*).

89. J. M. Thomas and R. Raja. Design of a "green" one-step catalytic production of ε-caprolactam (precursor of nylon-6), *PNAS*, **102**, 13732 (2005).

90. R. Raja, G. Sankar and J. M. Thomas. Bifunctional molecular sieve catalysts for the benign ammoximation of cyclohexanone: one-step, solvent-free production of oxine and caprolactam with a mixture of air and ammonia, *J. Am. Chem. Soc.*, **123**, 8153 (2001).

91. A. Uzun and B. C. Gates. Real-time characterization of formation and breakup of iridium clusters in highly dealuminated zeolite Y, *Angew. Chem. Int. Ed.*, **47**, 9245 (2008).

92. A. Uzun and B. C. Gates. Dynamic structural changes in a molecular zeolite-supported iridium catalyst for ethene hydrogenation, *J. Am. Chem. Soc.*, **131**, 15887 (2009).

93. A. Uzun, D. A. Dixon and B. C. Gates. Prototype supported metal cluster catalysts: Ir_4 and Ir_6, *Chem. Cat. Chem.*, **3**, 95 (2011).

94. M. Kim, M. Bertram, M. Pollmann, A. von Oertzen, A. S. Mikhailov, H. H. Rotermund and G. Ertl. Controlling chemical turbulence by

global delayed feedback: pattern formation in catalytic CO oxidation on Pt(110), *Science*, **292**, 1357 (2001).

95. J. Wolff, A. G. Papathanasiou, I. G. Kevrekidis, H. H. Rotermund and G. Ertl. Spatiotemporal addressing of surface activity, *Science*, **294**, 134 (2001).

96. P. L. Hansen, J. B. Wagner, S. Helveg, J. R. Rostrup-Melsen, B. S. Clausen and H. Topsøe. Atom-resolved imaging of dynamic shape changes in supported Cu nanocrystals, *Science*, **295**, 2053 (2002).

97. P. L. Gai. *In situ* environmental transmission electron microscopy, in *Nanocharacterization*, ed. A. I. Kirkland and J. L. Hutchison, RSC Publishing, Cambridge (2007), p. 268.

98. J. M. Thomas and P. L. Gai. Electron microscopy and the materials chemistry of solid catalysts, *Adv. Catal.*, **48**, 174 (2004).

99. P. L. Gai, C. C. Torandi and E. D. Boyes. *In situ* direct observation at atomic scale twinning transformations and the formation of carbon nanostructures in WC, in *Turning Points in Solid-State, Materials and Surface Science*, ed. K. D. M. Harris and P. P. Edwards, RSC Publishing, Cambridge (2008), p. 745.

100. J. M. Thomas and G. N. Greaves. Probing solid catalysts under operating conditions, *Science*, **265**, 1675 (1994).

101. S. R. Bare and T. Ressler. Characterization of catalysts in reactive atmospheres by X-ray absorption spectroscopy, *Adv. Catal.*, **52**, 339 (2009).

CHAPTER 5

SINGLE-SITE HETEROGENEOUS CATALYSTS FOR THE PRODUCTION OF PHARMACEUTICALS, AGROCHEMICALS, FINE AND BULK CHEMICALS

5.1 Introduction

The high-area, thermally stable nanoporous solids that we have described in earlier chapters make it possible, by judicious means, to "place" catalytically active centres in a spatially uniform fashion but so far apart (greater than ca 7 Å) from one another that each centre has the same energy of interaction between it and an incoming reactant. Moreover, because the open framework to which the active centres are attached is teeming with inner surfaces that line the nanopores, reactant species may gain ready access to, and quite bulky reaction products may freely diffuse away from, the energetically uniform single sites which are accommodated within these single-site heterogeneous catalysts (SSHCs).

In this chapter we cite numerous examples where SSHCs enable the production, often in a sustainable fashion, of a wide range of desired materials from fine chemicals — typified by vitamins, herbicides, fragrances and flavours — to bulk ones, such as benzaldehyde, adipic acid and nylon. Apart from providing economically attractive alternatives to current methods of production of these materials, SSHCs are also of great importance in facilitating green processes. Many of the examples we shall cite are solvent-free, use or generate no environmentally aggressive reagents or products, often entail single steps and may be so designed as to give rise to no unwanted salt or other products, thereby minimizing waste production.[1]

5.2 Fine Chemicals and Pharmaceuticals

The majority of the nanoporous SSHCs that my colleagues and I have used for these preparations are metal-(framework)-exchanged AlPOs consisting of redox, acidic or bifunctional active sites. Zeotype silica containing specific heteroatoms like Ti and Sn have also been used. Such nanoporous solids act, in effect, both as miniature reaction vessels and as solid (recyclable) heterogeneous catalysts.[2] In the examples to be described we shall be less concerned with the techniques of characterization of the catalyst than with the principles and the targets of the particular operation.

5.2.1 *Facile, one-step production of niacin (vitamin B₃) and other nitrogen-containing chemicals with SSHCs*

Niacin (3-picolinic acid, also known as vitamin B_3 or nicotinic acid), as well as other oxygenated products of 4-picoline, 4-methylquinoline and 4-methyl derivatives of pyrimidine, are of considerable value in pharmaceutical, agrochemical and fine-chemical applications. Both niacin (and nicotinamide, which we shall discuss further in Chapter 8) are building blocks for important co-enzymes required for the conversion of carbohydrates and for the metabolism of proteins and fats in the biological world. (The human body is incapable of producing niacin and is, therefore, dependent on intake through foodstuffs such as cereals to which this vitamin is deliberately added. Niacin is also used as a cholesterol-lowering agent.) Isonicotinic acid (4-picolinic acid) and its derivatives, especially isonicotinic acid hydrazide, have applications as antibacterial drugs for treating tuberculosis, psoriasis and arthritis and also as plant growth regulators.[3-5]

The industrial method of synthesizing nicotinic acid entails oxidizing nicotine with large quantities — about ninefold excess — of chromic acid, which, as seen in Scheme 5.1, generates an enormous quantity of by-products that are ecologically harmful: vast quantities of CO_2, NO_2 and also chromic oxide. Another industrial method is equally harmful: it entails the use of another strong, aggressive

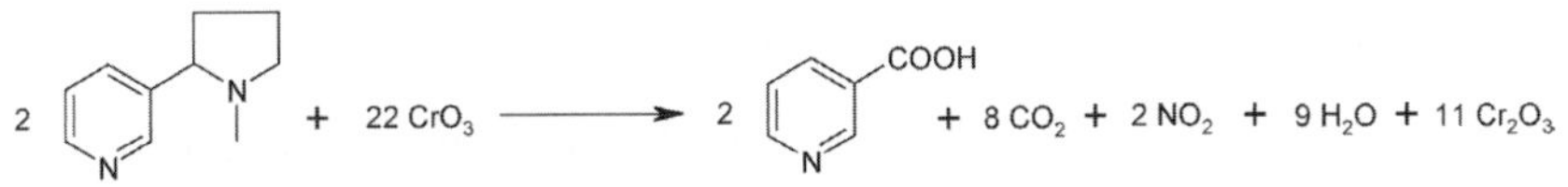

Scheme 5.1 Chromic acid oxidation of nicotine to nicotinic acid.

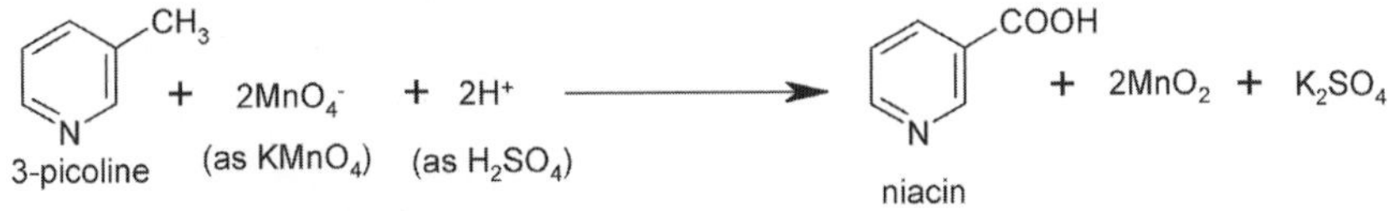

Scheme 5.2 Oxidation of 3-picoline with permanganates.

stoichiometric oxidant $KMnO_4$ (Scheme 5.2). The presence of small quantities of residual oxidant (chromium or manganese) in the product requires the use of expensive methods of purification, such as recrystallization, before the niacin can be used as a food additive.

A third method (not illustrated), which oxidizes nicotine in nitric acid, also gives rise to serious environmental problems.[6,7] And the liquid-phase oxidation of 3-picoline with homogeneous cobalt and manganese acetates and hydrobromic acid at *ca* 200°C and high pressures (*ca* 100 bar) does not give good yields or selectivities for the desired nicotinic acid.

The SSHC that my colleagues and I, in association with S. V. Ley, have designed for the purpose of preparing niacin involves the use of Mn^{III} AlPO-5 — see Figure 5.1, and compare with Figure 4.36. This schematic illustration shows the ease with which the reactants enter into the interior of the nanoporous catalyst.

For this purpose it is not O_2 or H_2O_2 that is used directly as the oxidant, but rather the soluble material acetylperoxyborate (APB), which when dissolved in aqueous solution in the presence of a single-site microporous catalyst,[8] such as Mn^{III} AlPO-5, yields a mixture of peroxyacetic acid (CH_3CO_3H) (PAA), acetic acid and hydrogen peroxide. The equilibrium is represented by $CH_3CO_2H + H_2O_2 \rightleftarrows CH_3CO_3H + H_2O$, and it has been shown experimentally by Harris and co-workers[8] that the ratio R of $[CH_3CO_3H][CH_3CO_2H]$ decays

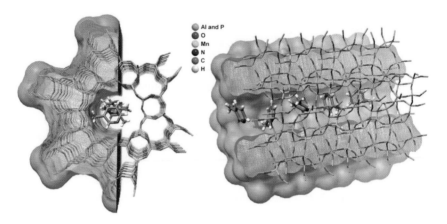

Figure 5.1 Top (left) and side (right) perspective views showing the accessibility of nitrogen-containing heterocyclic compound (3-picoline represented as example) to the internal, high-area, catalytically active surface of the single-site Mn AlPO-5 catalyst (the Mn^{III} active sites are shown in yellow). The computation of zeolitic surfaces was performed using a probe radius of 1.4 Å.[4]

gradually with time, as shown in Figure 5.2. Clearly, a decrease in the ratio R as the system approaches equilibrium leads to an increase in the concentration of H_2O_2. The latter liberates nascent oxygen at the active site (just as it does when brought into contact with TS-1, compare with Scheme 4.5 of Chapter 4). Although the kinetics of the selective oxidation of 3-picoline have been measured, the detailed mechanism has not been worked out. It is clear, however, (see Table 5.1) that the desired product (3-picolinic acid, nicotinic acid, niacin) is the overwhelmingly preferred product. What, in effect, the Mn^{III} AlPO-5 (SSHC) does is to favour the shape-selective conversion of 3-picoline to the required niacin. Unlike other means of oxidizing the reactant, here, little of the nicotinic acid *N*-oxide and only a comparatively small amount of 3-picoline *N*-oxide are formed — see Scheme 5.3.

We demonstrated that true shape selectivity is what governs the preferential production of vitamin B_3 (niacin, compound 2A in Scheme 5.3) by taking a pore-free, amorphous form of the Mn^{III} AlPO-5 catalyst in the oxidation of 3-picoline with APB. The results are summarized in Table 5.2.

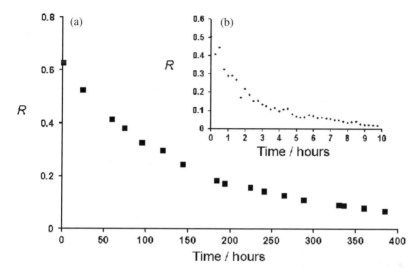

Figure 5.2 Plot of the ratio $R = [CH_3CO_3H]/[CH_3CO_2H]$, determined from solution ^{1}H NMR, as a function of time after dissolution of APB in water at (a) 25°C and (b) 65°C (see text).[8]

Table 5.1 Influence of temperature on the one-step, solvent-free production of niacin (nicotinic acid) using an SSHC: Mn^{III} AlPO-5.[a]

			Product Selectivity (mol %)			
T (K)	Oxidant	Conversion (mol %)	2A	3A	4A	Others
338	APB	37.0	78.5	21.4	—	—
358	APB	58.3	60.0	19.5	20.3	—
378	APB	86.9	42.6	15.0	27.8	14.5

[a] Reaction conditions: 3-picoline (2.8 g), catalyst (0.30 g) adamantine (internal standard, 0.5 g); time = 4 h. For the nature of the products, see Scheme 5.3.

5.2.2 Facile, one-step production of isonicotinic acid from 4-picoline

The hydrazide of isonicotinic acid is an important drug, used to counter tuberculosis, and is also utilized in the manufacture of agrochemicals and other pharmaceuticals (Scheme 5.4). As with 3-picoline (see Scheme 5.3) the challenge is to devise an oxidizing

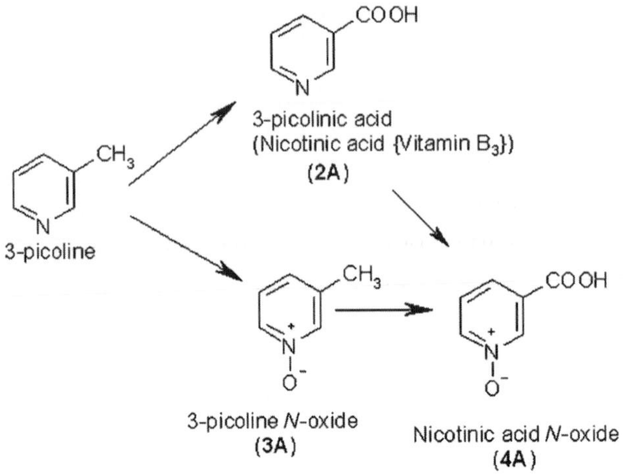

Scheme 5.3   The selective oxidation of 3-picoline to 3-picolinic acid, 3-picoline N-oxide and nicotinic acid N-oxide.[4]

Table 5.2   The influence of microporosity and redox-active site on the selectivity and activity in the oxidation of 3-picoline using APB.[4]

Oxidant	Reactant	Catalyst	T (K)	Conversion (mol %)	Products		
					2A	3A	4A
APB	3-Picoline	Microporous $Mn^{III}_{0.1}Al_{0.90}PO_4$	338	37.0	78.5	21.4	—
APB	3-Picoline	Amorphous $Mn^{III}_{0.1}Al_{0.90}PO_4$	338	7.4	—	41.5	50

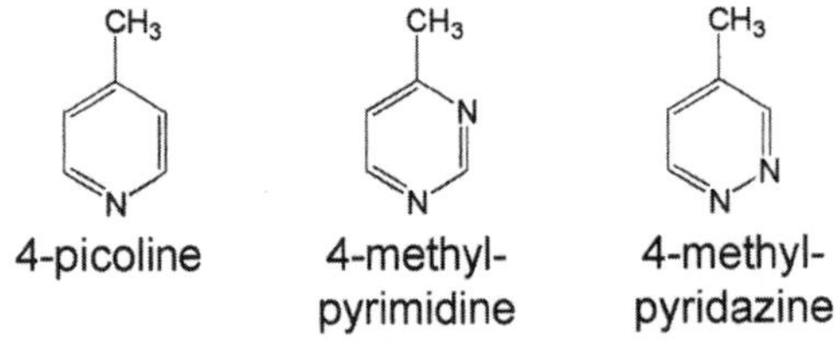

Scheme 5.4   4-Picoline; 4-methyl pyrimidine; and 4-methyl pyridazine.

process that converts the methyl group to the carboxylic acid, but leaves largely untouched the nitrogen in the aromatic ring. Again, the Mn^{III} AlPO-5 in conjunction with APB works effectively, in view of the shape selectivity that this combination imposes. At 368 K, for example, there is 72% conversion of the 4-picoline, 92% of the product being the desired isonicotinic acid (4-picolinic acid).[4]

5.2.3 *Production of pharmaceutically important derivatives of quinoline*[4]

Derivatives of quinoline carboxylic acids (see Scheme 5.5) are used extensively as biocides, pesticides, cancer drugs, seed disinfectants and especially antibacterial agents. The herbicides known as quinclorac and imazaquin, as well as plant growth regulators, antibiotics and antifungal agents for plants, are also derived from quinoline carboxylic acid, particularly cinchoninic acid.

Synthetically, the challenge again amounts to finding ways of selectively oxidizing the methyl group to the carboxylate while not generating either of the two *N*-oxides shown in Scheme 5.5. Almost all the industrial methods currently employed for this (and other) and related syntheses use environmentally aggressive dioxides and permanganates, and often two or more steps are required.

Again, Mn^{III} AlPO-5 in conjunction with APB proves effective, but higher temperatures are required (e.g. 47% conversion at 448 K) and the selectivity is lower than with the examples cited earlier

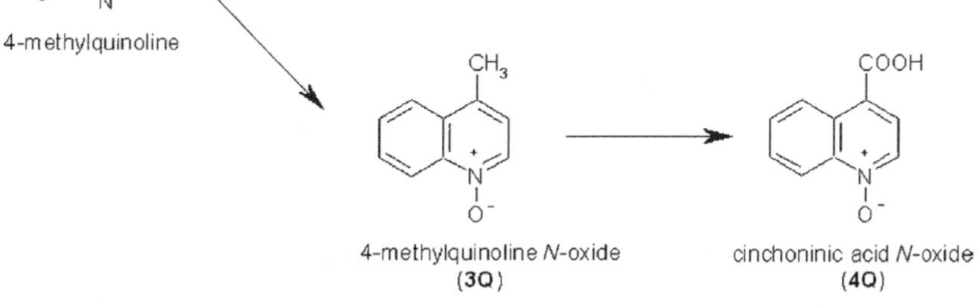

Scheme 5.5   Oxidation of 4-methylquinoline to cinchoninic acid.

(e.g. with the production of isonicotinic acid). The relatively large size of the 4-methylquinoline, in comparison to the 3- and 4-picolines, is doubtless responsible for the sluggish performance of this conversion, arising from the rather slow diffusion of the bulkier reactant.

Pyridazine carboxylic acids are used widely as plant growth regulators, a well-known example being Clofencet MON 21200, manufactured by Monsanto. 4-Methyl pyridazine carboxylic acid is produced with 100% selectivity (at 33% conversion of the 4-methyl pyridazine) at 338 K in 4 h — see Table 5 of Ref. 3. Peracetic acid or H_2O_2 as oxidants with the same Mn AlPO-5 catalyst are roughly comparable in the degree of conversion that they effect, but the selectivities in each case are much poorer, 35% and 41% respectively at 368 K.

5.3 Environmentally Benign Oxidative Methods of Producing Bulk Chemicals Using SSHCs

Almost all those chemicals used industrially as intermediates or end products — materials such as benzaldehyde, adipic acid and nylon — are prepared by oxidation using stoichiometric reagents like CrO_2Cl_2, HNO_3, CrO_3, $Na_2Cr_2O_7$, $KMnO_4$, SeO_2 and others as sources of oxygen. And even in standard textbooks of unimpeachable authority (young and old) reagents such as those enumerated along with other aggressive ones such as pyridinium dichromate acid are quoted as the reagents of choice for small-scale laboratory operations.[9,10] In an age when it is necessary, for ecological and other reasons, to use benign reagents like H_2O_2, or better still O_2 or air, as the oxidant it is necessary to design suitable solid catalysts that can enable oxidations to occur under mild conditions, and in such a manner as to enable the products to be separated readily from unconverted reagents. This is where the single-site heterogeneous catalysts described in earlier chapters come into their own. Moreover, many of them, as we describe in this chapter, can effect solvent-free and often single-step conversions with minimal or zero production of waste. It is salutary to recall how profligate many standard industrial methods of production really are. For example, in the oxidation of toluene to benzoic acid 1 kg of the (stoichiometric) oxidant, permanganate, generates 0.71 kg of MnO_2 as well as other salts.

In Sections 5.3.1 to 5.3.3 we illustrate the power of those SSHCs in which the active centres are of the redox kind, described and quantitatively analysed in Sections 4.7 and 4.8. In subsequent sections, we illustrate those SSHCs where the active centres are, first, of the Brønsted type (Section 5.4), followed by those where the active centres are Lewis acid centres (Section 5.5). Finally, we focus on the SSHCs where bifunctionality holds sway, as in the green method of preparing ε-caprolactam, the precursor of nylon 6.

5.3.1 *The synthesis of benzaldehyde from toluene*

5.3.1.1 *Background*

Benzaldehyde is a versatile material used extensively in the preparation of certain aniline dyes as well as perfumes and flavourings. In Europe, North America and Japan, as long ago as 1980, some 10×10^6 kg of benzaldehyde were produced annually, and the output has increased substantially, worldwide, ever since.

Commercially, benzaldehyde is produced by the chlorination of toluene; and the textbook description given by Fieser and Fieser[9] merits recollection here:

The chief technical process for production of benzaldehyde ... utilizes toluene as the starting material. One efficient method of conversion into the aldehyde is side-chain chlorination (preferably with illumination), and hydrolysis of the benzyl chloride cut. The hydrolysis is accompanied with water at 95 to 100°C in the presence of Fe powder or ferric benzoate as catalyst, lime is then added for neutralization and the benzaldehyde is steam-distilled. Benzoic acid usually appears as a by-product.

The same authors give details of another method of transforming toluene to benzaldehyde, using the classic Etard procedure:

A solution of two equivalents of CrO_2Cl_2 in CS_2 is added cautiously to the hydrocarbon (toluene) at 25–45°C. The red color of the reagent is discharged slowly and a chocolate-brown crystallizate separates consisting of a molecular complex containing two equivalents of the inorganic compound. The dry solid on treatment with water decomposes to give the aldehyde and an aqueous solution containing chromic acid and chromic chloride,[10–13] and the aldehyde must be removed rapidly by distillation or solvent extraction to avoid destruction:

$$ArCH_3 + 2CrO_2Cl_2 \rightarrow ArCH_3(CrO_2Cl_2)_2 \xrightarrow{H_2O} ArCHO$$

These passages, redolent of a bygone era in organic chemistry, emphasize how common it was to use reagents (like CrO_2Cl_2) and procedures of separation that are now strongly discouraged. They also betray an almost cavalier attitude, nowadays much frowned upon, towards the production of corrosive or toxic waste materials like chromic acid and chromic chloride.

Commercially, benzaldehyde is still largely produced by the chlorination of toluene,[13] followed by saponification but using procedures much streamlined from those employed in the classical era exemplified by Fieser and Fieser. Very recently, Kesavan *et al.*[14] have described a solvent-free oxidation of toluene in O_2 using nanoparticles of Au-Pd alloy supported on carbon or TiO_2. Their work is a laboratory-scale demonstration; and their procedures lead to high selectivities of other useful oxidation products such as benzyl benzoate (see Figure 5.3). These authors concluded that the high selectivity to benzyl benzoate **6** could result from four possible mechanisms:

- coupling of the aldehyde **3** and the alcohol **2** to give the hemiacetal **5**, followed by oxidation to the ester (see Scheme 1 of Figure 5.3);
- direct thermal (catalysed) dehydrative esterification between acid **4** and alcohol **2** (Scheme 1 of Figure 5.3);
- Cannizzaro reaction between alcohol **2** and aldehyde **3** via alkoxide **7** (Scheme 2 of Figure 5.3); and
- the so-called Tishchenko coupling of two aldehydes **3** via a "dimer" directly to the ester **6** (Scheme 3 of Figure 5.3).

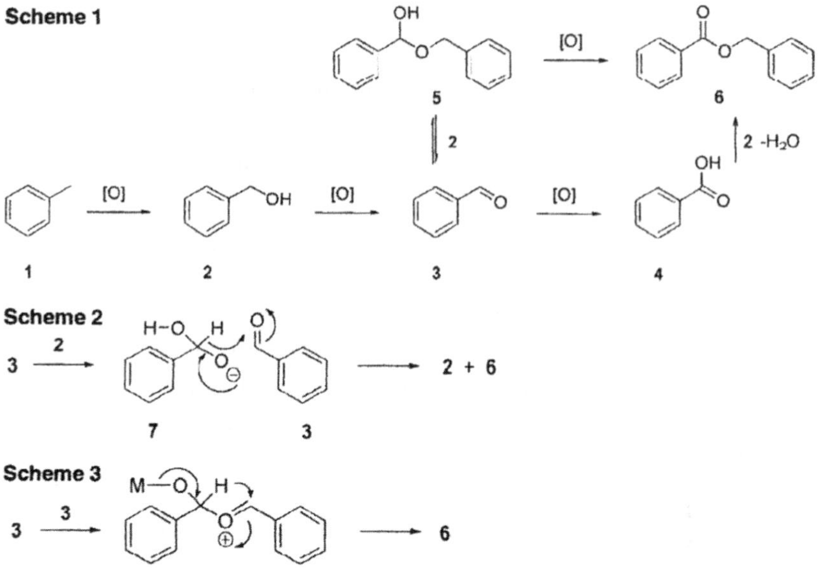

Scheme 1

Scheme 2

Scheme 3

Figure 5.3 Schemes presented by Kesavan *et al.*[14] in their discussion of solvent-free oxidation of toluene using Au-Pd alloy nanoparticles.

5.3.1.2 The solvent-free selective oxidation of toluene at a Mn^{III} redox centre[15]

We saw in Section 4.7 that single-site heterogeneous catalysts in which the active centre was an Mn^{III} (or Co^{III} or Fe^{III}) ion in a microporous oxide environment was extremely effective in activating C-H bonds in the presence of O$_2$, and, in Section 4.8, we gave a quantum chemical explanation for this remarkable selectivity. Based on prior work in the general field of oxyfunctionalization of hydrocarbons using SSHCs, my colleagues and I designed two specific types of solid catalysts, shown in Figure 5.4.

The type I catalysts are all framework-substituted transition-metal variants of AlPOs described earlier in this text. Type II catalysts are transition-metal-substituted phthalocyanines that, in their halogenated or nitrated states, are encapsulated within the cages of a faujasitic zeolite (typically Na$^+$-exchanged zeolite Y). This category of catalyst has also been shown[16-18] to effect the selective oxidation

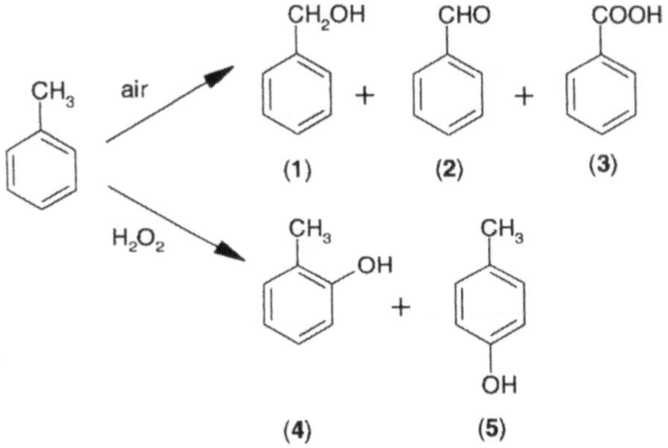

Figure 5.4 Two alternative ways of selectively oxidizing toluene.[15]

of a range of hydrocarbons using air, O_2, H_2O_2 and certain alkyl hydroperoxides. The aim of this work was to arrive at low-temperature, single-step processes, where the production of one or other of species **1** to **5** (see Figure 5.4) is maximized.

Within the type I category of SSHCs, two main structural types were employed:[15] M^{III} AlPO-5 and M^{III} AlPO-36 (where M = Co, Mn or Fe), the pore diameters being 7.3 Å for the MAPO-5 and 6.5 x 7.5 Å for MAPO-36. All the variants of the type I catalysts facilitate the selective, side-chain oxidation of toluene in O_2, the results being summarized in Table 5.3. It is seen that, depending upon the conditions used, it is possible to maximize the production of either benzaldehyde (by operating at a low temperature, 393 K) or benzoic acid (by operating at 423 K). In each case the percentage of benzyl alcohol also produced is either very small or non-existent.

5.3.2 *The one-step conversion of cyclohexane to adipic acid*

Already in Section 4.7 we have outlined both the environmentally unsatisfactory features of the present, popular method of oxidizing cyclohexane in air or O_2, via a mixture of cyclohexanol and cyclohexanone to adipic acid, and an alternative route using Fe^{III} AlPO-31 as an effective single-site heterogeneous catalyst.

Table 5.3   Catalytic oxidation of toluene[a] to benzaldehyde and benzoic acid.

SSH Catalyst	T (K)	Conversion (mol %)	Product Selectivity (mol %)[b]					
			1	2	3	4	5	6
$Mn_{0.04}Al_{0.96}PO\text{-}5$	393	7.3	3.7	78.8	16.1	—	—	0.5
$Co_{0.04}Al_{0.96}PO\text{-}36$	393	4.5	—	72.9	36.2	—	—	0.8
$Fe_{0.04}Al_{0.96}PO\text{-}5$	393	8.6	3.2	41.0	54.9	—	—	0.7
$Mn_{0.04}Al_{0.96}PO\text{-}5$	423	16.1	—	18.2	79.8	1.1	—	0.8
$Fe_{0.04}Al_{0.96}PO\text{-}5$	423	14.9	—	13.6	83.4	1.5	0.3	1.2
$Mn_{0.10}Al_{0.90}PO\text{-}5$	423	45.1	—	6.5	84.2	3.0	1.8	4.5

[a] Toluene = 25 g; catalyst = 0.75 g; air = 35 bar; time = 16 h.
[b] **1** = benzyl alcohol; **2** = benzaldehyde; **3** = benzoic acid; **4** = *o*-cresol; **5** = *p*-cresol; **6** = others (*m*-cresol, CO_2).

It is instructive to draw attention to the merits of APB again (compare with Section 4.7.2.1) as the oxidant for the preparation of adipic acid in an environmentally benign fashion. Using the same SSHC (namely Fe^{IV} AlPO-31) it is possible to convert over 85% of the cyclohexane, at 383 K, with a pH of 5.2 for the aqueous solution of APB, in 16 h to adipic acid with a selectivity of over 80%, the other products being cyclohexanone (mainly) and minor quantities of glutaric, succinic and valeric acids.[19]

Substituted phthalocyanines of Cu, Co and Fe, encapsulated in zeolites X and Y (like the type II category described in Figure 5.4), may also be used as single-site catalysts to effect benign oxidation of cyclohexane to cyclohexanol, cyclohexanone and adipic acid at ambient conditions using O_2 as well as alkyl (tertiary butyl, cyclohexyl and cumyl) hydroperoxides as the oxidants.[8] Rates of oxidation that are quite high (e.g. TOF = 400 h^{-1}) may be attained in this way. But it was found that solvents exert a major influence on product distribution.

Another aerobic route to adipic acid, starting from *n*-hexane and involving two spatially well-separated active sites in a microporous solid, was described in Section 4.7.2.2 — see Figure 4.38. But with the inevitable pressure in future to construct ways of using biological (plant) feedstock rather than fossil-based ones, much effort is now expended in devising efficient microbe-catalysed (i.e. biocatalytic) routes to prepare adipic acid. Frost and his colleagues[20] have, for example, demonstrated how plentifully available and renewable carbohydrates, such as D-glucose, offer potentially attractive ways to prepare adipic acid. We shall show, in Chapter 8, how nanocluster bimetallic catalysts, dispersed on mesoporous silica as a support, could play an important role in a sustainable method of preparing adipic acid. The single-site, multinuclear, bimetallic cluster $Ru_{10}Pt_2$, anchored on mesoporous silica, proved very efficient for the single-step hydrogenation of muconic acid (prepared biocatalytically) into adipic acid.[21]

Adipic acid, being the precursor to the synthesis of nylon 6,6 (which is a polymer of the dicarboxylic acid and hexamethylene diamine -$(CH_2)_6(NH_2)_2$), continues to be produced on a massive scale industrially. We have alluded earlier to the environmentally undesirable features of the current (nitric acid) method of producing adipic acid in a multi-step manner with the evolution of voluminous quantities of the greenhouse gas N_2O. In the year 2000 5% of the global emissions of N_2O came from adipic acid production plants. There is an urgent need to use the type of SSHCs that we describe in this monograph to replace the present method of industrial production.

5.3.3 *The one-step aerobic, solvent-free conversion of p-xylene to terephthalic acid*

The oxidant used in the so-called Amoco method of producing terephthalic acid is O_2, but the catalyst is a mixture of the acetates of Co and Mn, with some bromide ions present in the solution. It is used on a massive scale worldwide (driven mainly by the huge size of the polyethylene terephthalate (PET) polyester market). PET is formed from the condensation of terephthalic acid and ethylene glycol. The major disadvantage of the Amoco method, cited by Partenheimer,[13] is the necessity to use titanium-clad equipment to combat the corrosive nature of metal/bromine/acetic acid mixtures.

The approach[17] to the design by my group of suitable catalysts for the aerobic, bromine-free oxidation of *p*-xylene to terephthalic acid relies on the use of SSHCs in which the active centres are Co^{III} or Fe^{III} ions in framework sites substituting a small percentage of the Al^{III} ions in an appropriate AlPO microporous framework. These catalysts operate under solvent-free conditions without the need for corrosive activators and initiators such as bromine or bromide salts (see Figure 5.5). Their mode of operation[17] relies on shape-selective, free-radical processes of a spatially confined kind, where it was found that assembling up to 10 at % of transition metal ions in the framework of the microporous MAPO suppresses the formation of 4-formylbenzoic acid and increases the selectivity for terephthalic acid.

As in the case of Co^{III} AlPO-18, containing *ca* 10 at % of Co^{III} ions in the framework,[22] one might envisage the oxyfunctionalization occurring simultaneously at both methyl ends of *p*-xylene, which could well be the reason for the higher selectivity of the Mn^{III}

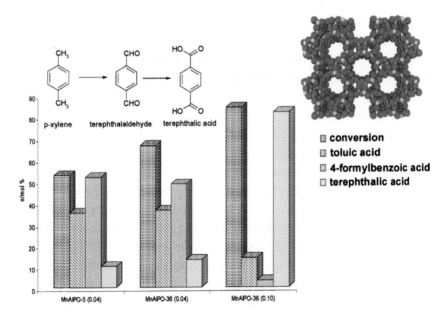

Figure 5.5 Bromine-free route for the aerobic oxidation of *p*-xylene to terephthalic acid.

AlPO-36 (0.10) catalyst for terephthalic acid (Figure 5.5), compared to Mn^{III} AlPO-36 (0.04). Contrary to earlier observations,[23] where it was shown that this reaction proceeds in a sequential manner via the formation of *p*-tolualdehyde and *p*-toluic acid, thereby requiring the use of corrosive bromide activators (for activating the second methyl group), our detailed kinetic experiments have revealed the simultaneous oxidation of both methyl groups in *p*-xylene to produce terephthaldehyde and terephthalic acid, as shown in Figure 5.5.

5.4 Environmentally Benign, Brønsted Acid-catalysed Production of Bulk Chemicals with Microporous SSHCs

Brønsted acid catalysts have been used on an industrial scale for several decades, but the need to develop new ones has recently intensified in order to minimize the disadvantages associated with the present extensive use of corrosive and environmentally dangerous liquids such as concentrated phosphoric, sulfuric and hydrofluoric acids.[24–26] For example, over 15 million tons of sulfuric acid are consumed annually for the production of industrially important chemicals.[26]

Very strong acids — capable of catalysing demanding reactions such as alkylations of isobutene by butane to yield high-octane fuels — like the liquids mentioned earlier are often supported on siliceous or clay minerals. But it is urgently necessary to develop new solid catalysts that do not suffer from the environmental disadvantages and adverse corrosive properties of liquids.

In Chapter 4 several examples of Brønsted-acid-catalysed reactions (of relevance to the production of bulk chemicals) were outlined, one of the most important being the MTO and MTH reactions, whereby methanol is converted to light olefins and other hydrocarbons (see Section 4.4). Further examples in which SSHCs of a microporous kind are used are given in Table 5.4.[27]

As described in Chapter 4, as well as the pentasils (ZSM-5 and ZSM-11), a host of other aluminosilicate Brønsted acid catalysts, such as zeolite Beta, ferrierite, mordenite and erionite, in addition to SAPOs and MAPOs, constitute prominent examples of SSHCs for the production of bulk chemicals.

Table 5.4   Some economically significant conversions effected by SSHCs of a Brønsted acid kind.

Alkylation of benzene with ethene (e.g. Mobil–Badger process for producing ethyl benzene using ZSM-5)
Xylenes from the isomerization and disproportionation of toluene
Oligomerization of olefins
Cumene from benzene plus propene
Isomerization of methyl naphthalenes
Detergent alkylate from benzene plus C_6 to C_{14} olefins
Methanol to olefins and methanol to hydrocarbons

The recent drive to convert biomass to such commodities as transportation fuels and a range of other desirable products has involved intensive research for new microporous solid acid catalysts — see, in particular, Refs 28–30 and Appendix I.

5.5 Transformations Involving Lewis Acid Microporous Catalysts

In Chapter 4 we outlined a number of the important bulk chemical processes involving TS-1, which is silicalite endowed with Ti Lewis acid centres (see Section 4.5 and Scheme 4.5). With the increasing interest in catalytic conversions of carbohydrate and other sustainable feedstocks, there is now much interest in microporous solids in which other Lewis acid centres, notably Sn and Zr, are present as single sites.[31–33] Moreover, Corma and co-workers,[34,35] as we shall discuss shortly, have demonstrated the value of Sn Lewis acid centres to effect Baeyer–Villiger reactions.[36–38] We now proceed to describe the salient features of these Lewis acid catalysts.

5.5.1 *Conversions of sugars to lactic acid derivatives using Sn-based zeotypic SSHCs*

Lactic acid (α-hydroxypropionic acid) is an important chemical that is used for production of biodegradable polymers and solvents; and there is a thriving industry involving the production of polylactic acid (PLA) polymers from corn and other renewable sources. The technical textiles derived from PLA consist of composites, medical appliances, filtration fabrics, packaging material and many other industrial fabrics.

The industrial production of lactic acid is based on the anaerobic conversion of glucose and sucrose using microbial fermentation.[31] The major complications associated with this process are the need to neutralize lactic acid with a stoichiometric amount of base during the fermentation process and the energy-intensive work-up of lactic acid from the aqueous fermentation broth.

However, a different route to lactic acid, based on the isomerization of the triose sugars dihydroxyacetone (DHA, a keto-monosaccharide containing three carbon atoms) and glyceraldehyde (GLA, which is an aldotriose) — see Figure 5.6 — may be envisaged, as reported by Taarning *et al.*[31] Such a process could be based on glycerol as a raw material, which can be converted to DHA and GLA by catalytic oxidation or by fermentation.[39]

Recognizing that very few compounds of major commercial interest are currently accessible from carbohydrates by non-fermentative procedures, Holm *et al.*[32] have recently described a catalytic process, involving a strong Lewis acid SSHC, for the direct formation of methyl lactate from common sugars.

It is relevant to recall that carbohydrates represent the largest fraction of biomass, and several strategies for their efficient use as a commercial chemical feedstock have already been explored, with a view, ultimately, of replacing petroleum.[31–33,40a,41] Because the thermal instability of carbohydrates is a significant obstacle in this regard, biochemical processes have hitherto proven to be more applicable than catalytic ones, chiefly because they can be operated at low temperatures. And, to date, relatively few chemical products are directly obtainable from carbohydrates by using heterogeneous catalysis. To be sure, gluconic acid and sorbitol can be obtained from

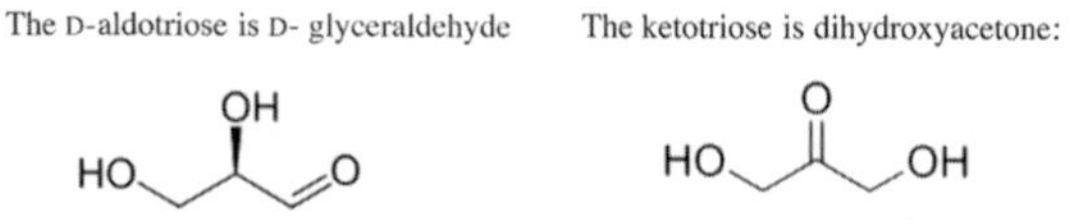

The D-aldotriose is D- glyceraldehyde   The ketotriose is dihydroxyacetone:

Figure 5.6 Two important triose monosaccharides: (left) D-aldotriose (D -glyceraldehyde); (right) ketotriose (dihydroxyacetone) (see text).

glucose by oxidation and hydrogenation, respectively; and acidic catalysts, as shown by Dumesic and co-workers, are capable of producing the desirable product, 5-hydroxymethylfurfural (HMF).[30,41]

Returning to the relevance of lactic acid in a sustainable world, not only is it, as outlined earlier, capable of leading to a wide range of biodegradable polymers, it is also potentially a precursor to the production of a variety of other useful compounds through catalytic transformation.[42,43]

Holm *et al.*[32] have reported that fructose may be readily converted to methyl lactate in yields of up to 44%. They argued — see Figure 5.7 — that this reaction proceeds via a so-called retro-aldol reaction of fructose to form the two trioses shown in Figure 5.6. These trioses are readily converted to the thermodynamically very stable methyl lactate through sequential dehydration and methanol addition, followed by a 1,2-hydride shift, as reported by Taarning *et al.*[31]

Using a range of Lewis acid SSHCs composed of zeotype Beta in which Ti, Zr and Sn single sites were separately incorporated into the siliceous Beta microporous framework with an Si/M ratio of 125 (where M = Ti, Zr or Sn), Holm *et al.*[32] found conversions of 98% and yields ranging from 31% and 36% for Ti-Beta, of 33% for Zr-Beta and 44% and 64% (in the conversion of fructose and glucose respectively) for Sn-Beta at a temperature of 160°C. This sequence of

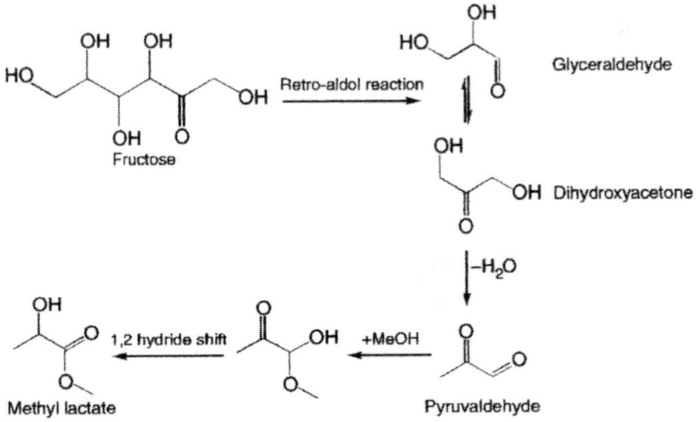

Figure 5.7 Proposed reaction pathway for the conversion of fructose to methyl lactate. The reaction formally comprises a so-called retro-aldol fragmentation of fructose and isomerization-esterification of the trioses.[32]

catalytic activity (Sn > Zr > Ti) scales with the strength of the Lewis acidity of the respective SSHCs,[31] as determined spectroscopically.

5.5.2 *Single-site, Lewis acid microporous catalysts for the isomerization of glucose: a new efficient route to the production of high-fructose corn syrup*[33]

The isomerization of glucose into fructose is a reaction conducted on a massive scale industrially as may be seen from the fact that high-fructose corn syrup (HFCS) has reached a global production that exceeds 8×10^6 tons per annum. (In the US alone, the per capita consumption of HFCS was 37.8 lb per annum in 2008.) Currently the process is catalysed enzymatically. Very recently Moliner *et al.*[33] have shown that the zeotypic Lewis catalyst, Sn-Beta, is highly active in the isomerization of glucose in water; moreover, this inorganic SSHC resembles the performance of enzymatic catalysts by generating remarkably high fructose yields at glucose conversions near the reaction equilibrium. Furthermore, this catalyst, unlike enzymatic catalysts, maintains high activity over multiple cycles. In addition, it can be readily regenerated with a mild calcination process and operated over a range of temperatures, at which the enzyme becomes unstable, and the Sn-Beta catalysts can also function effectively in a highly acidic environment, making them attractive candidates for sequential or one-pot catalysed (cascade) reaction sequences (see Scheme 5.6).

The Sn-Beta Lewis catalyst (Si:Sn ≈ 50:1) described by Moliner *et al.*[33] is also capable of operating in such a manner as to couple isomerizations with other acid-catalysed reactions, including hydrolysis/isomerization or isomerization/dehydration reaction sequences, involving the sequence of starch to fructose to 5-hydroxymethylfurfural. On catalytically hydrating HMF it is converted[30] to levulinic acid $COOH(CH_2)_2COOH$, that constitutes a viable substitute for adipic acid in many polyester products.

Detailed mechanisms of these conversions have not yet been worked out, but there is no doubt concerning the actual nature of the Lewis acid active site.[34,35] The Sn in the siliceous framework of the Beta structure is in four-coordination, best represented by $(-Si-O-)_3Sn-OH$.

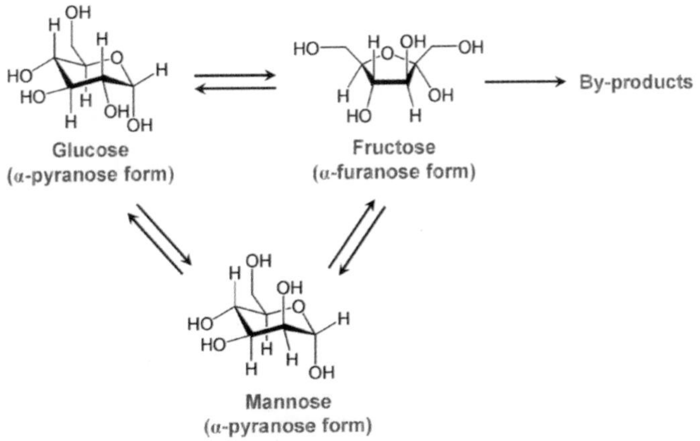

Scheme 5.6 Representation of the glucose isomerization reaction pathways cata-lysed by either biological or chemical catalysts (see text).

We shall see in the next section that this active site is especially good in the catalytic conversion of ketones to lactones (the Baeyer–Villiger reaction[36,44]).

5.6 Baeyer–Villiger Oxidations of Ketones to Lactones with SSHCs

5.6.1 *Introduction*

The Baeyer–Villiger oxidation is a reaction of major synthetic inter-est in organic chemistry with a large range of applications spanning such varied areas as the synthesis of antibiotics and steroids, the synthesis of pheromones for agrochemistry and the synthesis of a diverse range of monomers for the polymer industry. Despite its longevity,[44] it continues to elicit a great deal of interest and serves as a stepping stone for many commercially significant products.

For most of its long period of use since its discovery over a cen-tury ago, the reaction has been carried out stoichiometrically. The original oxidant was Caro's acid, monopersulfuric acid H_2SO_5, and indeed Baeyer and Villiger in 1899 used this reagent to oxidize men-thone and tetrahydrocarvone to the corresponding lactones (see Scheme 5.7). These substances are still relevant to the flavour and fragrance industries.

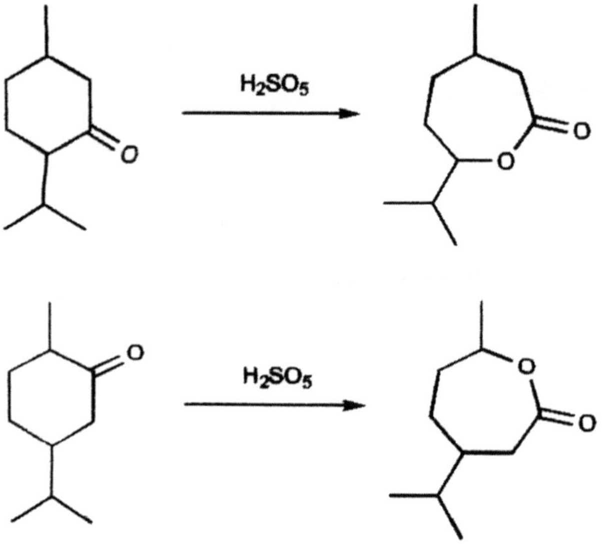

Scheme 5.7 Caro's acid, H_2SO_5, was used by Baeyer and Villiger[44] to oxidize menthone and tetrahydrocarvone into products that are used in flavours and fragrances.

A wide variety of other stoichiometric oxidants may be used in this reaction, typically CrO_3, $KMnO_4$, $Pb(OAc)_4$, RuO_4 and Ag_2O, almost all of which, in the context of green chemistry and clean technology, are now unacceptable. Consequently, much effort has recently been spent on developing appropriate, benign catalysts for effecting Baeyer–Villiger reactions. We discuss two main types here, each of them involving single-site solid catalysts. One utilizes molecular oxygen, the other H_2O_2.

5.6.2 A redox SSHC for Baeyer–Villiger aerobic oxidations under Mukaiyama conditions[38,46]

In the early 1990s, Mukaiyama and co-workers described the use of sacrificial aldehydes for the aerobic epoxidation of olefins in the presence of Ni complexes (in solution). Their procedure exploits the facile autoxidation of aldehydes to promote the *in situ* formation of peroxy acids; and this approach is highly successful in promoting the aerobic oxidation of a variety of substrates including olefins, aldehydes and lactams.[46] My colleagues and I showed[47] that it could be readily adapted to work heterogeneously, using a single-site (redox) MAPO catalyst for the low-temperature epoxidation of alkenes using O_2 or air as oxidant.

It is widely acknowledged[36] that the catalyst in the Mukaiyama procedure serves merely to promote the easy autoxidation of benzaldehyde to produce perbenzoic acid (or its peroxy radical) in a manner almost identical to the well-known industrial synthesis of acetic acid.[48] Using our know-how concerning high-oxidation-state transition metal ions framework-substituted into AlPOs, in particular Mn^{III} or Co^{III} ions, we developed[37,38] a series of effective SSHCs for the Baeyer–Villiger oxidations, with high lactone selectivity ranging from 75% to 98%, and conversion from 65% to 87%, when the following (redox) single-site catalysts were used (at 323 K, contact time 6 h): Mn^{III} AlPO-5, Mn^{III} AlPO-36, Co^{III} AlPO-5 and Co^{III} AlPO-36. The substrates (ketones) converted to their corresponding lactones were cyclohexanone, cyclopentanone, 2-methylcyclohexanone and adamant-2-one (see Table 5 of Ref. 37 for detailed experimental conditions).

So, in essence, the sacrificial benzaldehyde fed into the reaction sphere permits aerobic Baeyer–Villiger oxidations to occur under mild, low-temperature (environmentally acceptable) conditions.

Table 5.5 Baeyer–Villiger oxidation of dihydrocarvone with different catalysts, using H_2O_2 and peracid as oxidant.

Oxidant	Reactant Conversion (%)	Products Selectivity (%)		
Sn-Beta/H_2O_2	68	100	0	0
MCPBA[a]	85	11	71	18
Ti-Beta/H_2O_2	46	0	79[b]	0
MTO/H_2O_2[c]	9	30	33	20

[a] Metachloroperbenzoic acid.
[b] The missing 21% comprising products from ring opening of the epoxide.
[c] Methyltrioxorhenium.

There is a kinship between the role of Mn^{III} (and Co^{III}) oxide-centred ions here and the redox facility of these two ions as they operate in the oxyfunctionalization of alkanes, described in detail in Chapter 4. There is the necessity of requiring two oxidation states of the transition metal, and free radicals participate in the selective oxidation:

$$PhCHO + Mn^{III} \cdot PhCO \cdot + H^+ + Mn^{II}$$
$$PhCO + O_2 \rightarrow PhCOOO$$
$$PhCOOO \cdot + PhCHO \rightarrow PhCOOOH + PhCO \cdot, \text{ etc.}$$

This sequence of steps (described fully in Ref. 37) is also involved in the catalytic selective oxidation of a number of alkenes in which high yields of the epoxides of the following reactants occur[37,47]: α-(+)-pinene; R-(+)-limonene; styrol; 1-hexene; and cyclohexene.

Most of these produced the epoxides in high yields, but some diols were also generated in the case of cyclohexene.[47]

5.6.3 Sn-centred single-site microporous catalysts for Baeyer–Villiger oxidations with H_2O_2

Corma *et al.*[34,35,49] have used a Lewis acid SSHC for this reaction. It entails the siliceous (Al-free) form of the zeotype Beta framework in which isolated, tetrahedrally coordinated Sn ions are embedded in a spatially separated fashion into the framework of this large-pore solid. Their catalyst exhibits good activity and high selectivity to the corresponding lactone.[34] In addition, they showed that when a double bond is also present in the reactant (cyclic) ketone, a high chemoselectivity is observed, as depicted in Table 5.5.

By carrying out a mechanistic study using methylcyclohexanone labelled with ^{18}O as reactant, it was concluded[35] that the Baeyer–Villiger oxidation, with H_2O_2 at a Sn-Beta single-site catalyst, involves a so-called Criegee adduct — named after the German scientist who first proposed and identified it. In this adduct, the H_2O_2 is bound to the Sn atom via the oxygen of the ketone. The complete (proposed) mechanism is schematized in Figure 5.8.

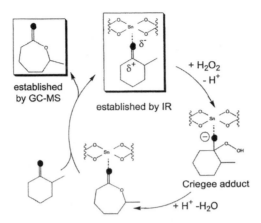

Figure 5.8   Mechanism of the Baeyer–Villiger oxidation of 2-methylcyclohexanone catalysed by Sn-Beta.[35]

5.7 The Crucial Role of Single-site Microporous Catalysts in New Methods of Synthesizing ε-Caprolactam and Nylon 6

5.7.1 *Introduction*

Nylon comes in many different forms. The first kind of nylon, produced by Carothers at the DuPont Company in 1939, is nylon 6,6. This is a condensation product of adipic acid and hexamethylenediamine (six atoms in the acid and six in the backbone of the diamine), and it is made up of polyamide chains. Different pairs of diacids and diamines yield new variants of nylon. Thus nylon 5,10, which is formed from sebacic acid and pentamethylenediamine, is one example; others are designated 6,10; 6,12; etc. The form of nylon that is of greatest interest commercially, for reasons that are explained later, is nylon 6, which was first prepared by Schlack at the I.G. Farben Co. Nylon 6 is the polymeric form of ε-caprolactam (see Figure 4.40 and other representations later in this chapter) — see Scheme 5.8.

5.7.2 *The primacy of nylon 6*

Like many of the other forms of nylon, it possesses a range of commercially attractive properties that make it extremely important as a

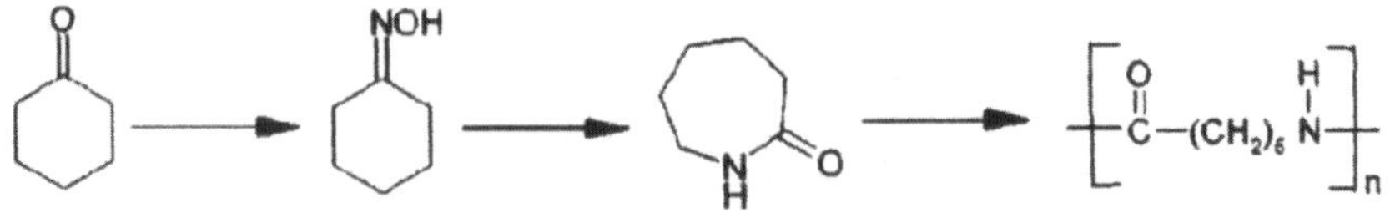

Scheme 5.8 Summary of sequence of conversions that produce ε-caprolactam (and its linear polymer, nylon 6) from cyclohexanone (**1**) via its oxime (**2**).

Table 5.6 Important practical properties of nylon 6.

1. Nylon 6 fibres are tough, possessing high tensile strength as well as elasticity and lustre.

2. High stability in O_2: does not degrade readily.

3. Resistant to abrasion; readily made into any desired shape; readily sterilized.

4. Lightweight: low aptitude to absorb water.

5. Electrically insulating; durable; adheres to rubber readily.

6. May be manufactured in many colours because of ease with which it can be dyed.

7. Softening temperature may be significantly increased by incorporation of thin lamellae of montmorillonite clay to form nanocomposites usable as components of car engines.[a] Nanocomposites with carbon nanotubes also feasible.

[a] N. Hasegawa, H. Okamoto, M. Kato, A. Usuki and N. Sato, *Polymer*, **44**, 2933 (2003).

synthetic material; these are summarized in Table 5.6. But over and above all these useful properties, many of which are shared by other forms of nylon, it is readily degraded (recycled) using superheated steam and other methods. Unlike other nylons which, after use, end up in landfills, nylon 6 (polycaprolactam) can be reprocessed and broken down to its individual monomer units. For this reason alone it is an ideal example of an environmentally friendly synthetic fibre.

In view of this collection of good qualities, nylon 6 is used extensively, e.g. as thread in bristles for toothbrushes, surgical sutures and strings for acoustic and classical musical instruments, including guitars, violins and cellos. It also finds extensive application in ropes, threads, nets, filaments as well as hosiery and knitted garments.

To appreciate the industrial magnitude of nylon 6 manufacture, it is relevant to recall that, at one particular plant alone (operated by the Honeywell Co.), 350,000 metric tons of caprolactam are produced annually.

5.7.3   *Existing routes to the synthesis of ε-caprolactam*

There are many ways in which the essential intermediate, cyclohexanone oxime (which on isomerization yields ε-caprolactam), may be formed. Some start from cyclohexane; usually, however, the starting point is cyclohexanone — see Figure 5.9, which indicates the routes favoured by the manufacturers therein named. Other methods of producing the key oxime and caprolactam are still being explored, and good progress has been made using Au nanoparticles supported on TiO_2 for either oxidation (of cyclohexylamine) or

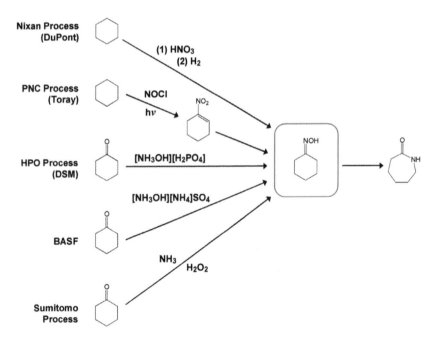

Figure 5.9   A summary of some of the methods feasible for the production of the oxime of cyclohexanone for subsequent preparation of caprolactam.

reduction (of nitrocyclohexane) to yield cyclohexanone oxime. There are also other routes, rather less attractive since they involve multiple steps, of reaching the key oxime or the ε-caprolactam. Hydrocyanation and hydrogenation to produce first a linear intermediate then cyclization to the caprolactam is a viable one.

Current production of ε-caprolactam relies on two main, especially popular processes: the Raschig one (Scheme 5.9), and the so-called Sumitomo/EniChem process (Scheme 5.10). As mentioned earlier, very large quantities of $(NH_4)_2SO_4$ (a low-grade fertilizer) are produced in the Raschig process, that utilizes hydroxylamine hydrosulfate to ammoximize (form the oxime of) the cyclohexanone. Oleum is then used to isomerize the oxime (by Beckmann rearrangement) to the caprolactam. In the Sumitomo/EniChem process NH_3 and H_2O_2 convert the cyclohexanone, in the presence

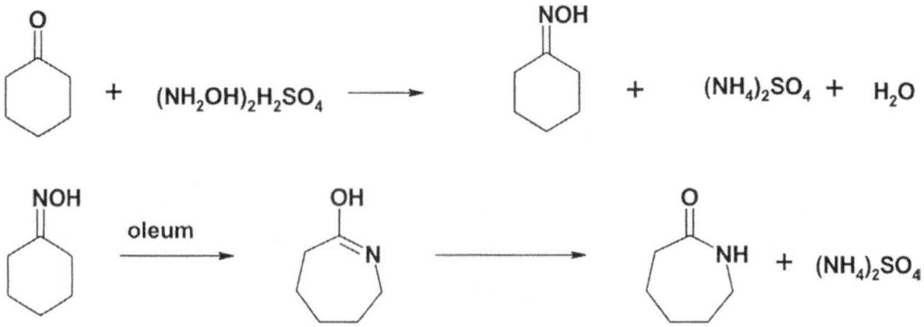

Scheme 5.9   The Raschig process for the production of ε-caprolactam.

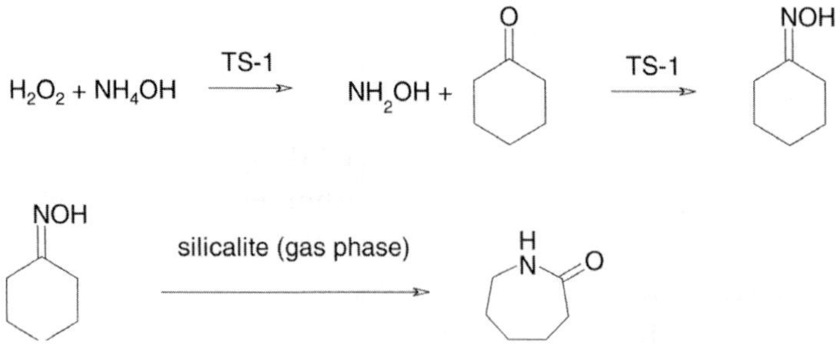

Scheme 5.10   The EniChem/Sumitomo method of synthesizing ε-caprolactam using exclusively zeotypic (microporous) SSHCs.

of a single-site (TS-1) catalyst, to the imine and then to the oxime, which in turn undergoes the Beckmann rearrangement in the gas phase over a weakly acidic silicalite catalyst to yield the caprolactam.

Other variants of the Sumitomo/EniChem process continue to be described — see, for example, Saxena *et al.*[51] who also use TS-1 as a catalyst but the urea hydrogen peroxide (UHP) adduct to yield both the imine and the oxime.

5.7.4 *The design of a green, one-step production of ε-caprolactam using a bifunctional SSHC*

In 2005, the author and his colleagues,[52] building on earlier work using nanoporous bifunctional SSHC,[53] demonstrated a route to cyclohexanone oxime — that did not generate unwanted ammonium sulfate — which was environmentally benign, used air (or O_2) as the oxidant and was solvent-free. Their catalysts were bifunctional because they contained both Brønsted acidic and redox single sites. The Brønsted acidity arose because ions such as Mg^{II} and Zn^{II} replaced Al^{III} ions in AlPO frameworks; and the redox sites were Mn^{III}, Co^{III} or Fe^{III} ions substituting Al^{III} ions in the solids. At these redox sites, NH_3 in the presence of air or O_2 forms hydroxylamine *in situ*. Moreover, because the open structure of these bifunctional catalysts (typically Co^{III} or Mg^{II} AlPO-5) contains pores large enough to facilitate the diffusion of reactants, intermediates and products within them, the parent ketone is sequentially and smoothly converted to the lactam when it, air and NH_3 are brought into contact with one another within the solid catalysts. Figure 5.10 summarizes both the nature and the practical value of this family of bifunctional catalysts.

The solids that Thomas and Raja[53] designed for this purpose, like those that they used to effect other environmentally benign oxidations (such as the conversion of cyclohexane in air to adipic acid — see Section 5.3), are classic examples of single-site heterogeneous catalysts that are the principal theme of this monograph: Tables 5.7 and 5.8 summarize the results obtained with the family of bifunctional catalysts designed by them.

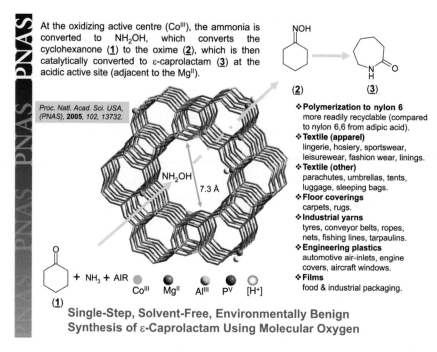

At the oxidizing active centre (Co^{III}), the ammonia is converted to NH$_2$OH, which converts the cyclohexanone (**1**) to the oxime (**2**), which is then catalytically converted to ε-caprolactam (**3**) at the acidic active site (adjacent to the Mg^{II}).

Proc. Natl. Acad. Sci. USA,
(PNAS), **2005**, *102, 13732.*

NH$_2$OH

7.3 Å

+ NH$_3$ + AIR

Co^{III} Mg^{II} Al^{IIII} P^V [H$^+$]

(**1**)

(**2**)   (**3**)

❖ **Polymerization to nylon 6**
more readily recyclable (compared to nylon 6,6 from adipic acid).
❖ **Textile (apparel)**
lingerie, hosiery, sportswear, leisurewear, fashion wear, linings.
❖ **Textile (other)**
parachutes, umbrellas, tents, luggage, sleeping bags.
❖ **Floor coverings**
carpets, rugs.
❖ **Industrial yarns**
tyres, conveyor belts, ropes, nets, fishing lines, tarpaulins.
❖ **Engineering plastics**
automotive air-inlets, engine covers, aircraft windows.
❖ **Films**
food & industrial packaging.

Single-Step, Solvent-Free, Environmentally Benign Synthesis of ε-Caprolactam Using Molecular Oxygen

Figure 5.10   Simplified illustration of the single-step, solvent-free, environmentally benign synthesis of ε-caprolactam using molecular O$_2$.[53]

Table 5.7 A summary illustrating the effectiveness of bifunctional, AlPO-5 derived SSHCs for the solvent-free production of ε-caprolactam (derived from Table 1 of Ref. 53).

Composition of Bifunctional Catalyst	Conversion (mol %)	Product Selectivity (mol %)		
		Oxime	ε-Caprolactam	Others
$Mn_{0.02}Mg_{0.02}Al_{0.96}PO_4$	68.3	5.7	77.9	16.4
$Fe_{0.02}Mg_{0.02}Al_{0.96}PO_4$	71.9	11.5	72.0	16.3
$Mn_{0.02}Zn_{0.02}Al_{0.96}PO_4$	61.5	6.3	65.0	28.6

Conditions: Catalyst, 0.50 g; cyclohexanone, 5.0 g; air, 35 bar; NH$_3$, 14.6 g; temperature, 353 K, time = 8 h.

Others = High-molecular-weight conjugated products through aldol condensation of cyclohexanone.

Table 5.8 Catalytic low-temperature, solvent-free ammoximation of cyclohexanone: Performance of three bifunctional catalysts in which the Brønsted acid centre is kept constant.

Microporous Bifunctional Catalyst	Conversion (mol %)	Product Selectivity (mol %)		
		Oxime	ε-Caprolactam	Others
$Mn_{0.02}Si_{0.02}Al_{0.96}PO_4$	76.5	86.6	3.3	10.0
$Co_{0.02}Si_{0.02}Al_{0.96}PO_4$	68.2	79.6	4.7	14.7
$Fe_{0.02}Si_{0.02}Al_{0.96}PO_4$	87.5	63.2	5.5	31.2

See Table 5.7 for conditions.

5.7.5 *Optimizing SSHCs for oxime production*

Whereas the proof of principle contained in the work described in Section 5.7.4 for the one-pot synthesis of caprolactam constitutes a significant advance in catalyst design, this work also opens the way to optimizing the catalytic production of oximes, which are extremely important intermediates in the synthesis of new materials. Raja and co-workers[54] began in 2009 a systematic study of fine-tuning the nature and extent of the single sites introduced into the AlPO-5 framework. Rather than tackle the problem of optimizing the nature of M^{II} and M^{III} ions so as to achieve the best efficiency in producing the caprolactam, it was felt that the more limited but important target of maximizing the production of oxime (from cyclohexanone, air and NH_3) was a desirable end. This work continues apace; and significant advances have already been made, as shown by the selectivities, conversions and turnover numbers achieved with the bisubstituted AlPO (namely CoTi AlPO-5) shown in Figure 5.11. For reasons that are somewhat elusive, it seems that a synergy exists between Co^{III} and Ti^{IV} ions in the AlPO-5 framework in the ammoximation of cyclohexanone. This bisubstituted SSHC surpasses TAPO-5 (i.e. Ti AlPO-5), and far surpasses TS-1 in the catalytic ammoximation of cyclohexanone.

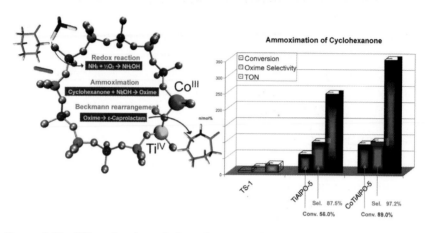

Figure 5.11 Bifunctional catalysis and consecutive reactions. A new single-site (synergistic) catalytic process for the ammoximation of ketones.

5.8 Envoi

This chapter highlights the tangible way in which the principles enumerated in earlier sections of the monograph pertaining to single-site heterogeneous catalysts of an ordered, microporous nature lead to a wealth of methods for the synthesis of a wide range of useful products. Almost invariably these new syntheses are of a green character, often involving solvent-free, single-step and sustainable processes. Moreover, they are environmentally benign and involve little or no waste.[55] Apart from describing viable methods of synthesizing vitamin B_3 and other pharmaceuticals, we also show how bulk (commodity) chemicals like adipic, terephthalic and lactic acids, high-fructose corn syrup, benzaldehyde, caprolactam and nylon 6 may be cleanly prepared. We also discuss and illustrate modern means of effecting the Baeyer–Villiger reaction that converts cyclic ketones to (desirable) lactones, products that are extensively used as flavours and fragrances.

References

1. J. M. Thomas and R. Raja. Innovations in oxidation catalysis leading to a sustainable society, *Catal. Today*, **117**, 22 (2006).

2. R. Raja and J. M. Thomas. Nanoporous solids as receptacles and catalysts for unusual conversions of organic compounds, *Solid State Sci.*, **8**, 326 (2006).

3. J. M. Thomas and R. Raja. Designed open-structure heterogeneous catalysts for the synthesis of fine chemicals and pharmaceuticals, in *Zeolites to Porous MOF materials — The 40th Anniversary of International Zeolite Conferences*, ed. R. Xu, Z. Gao, J. Chen and W. Yan, Elsevier, Amsterdam (2007), p.19.

4. R. Raja and J. M. Thomas, M. Greenhill-Hooper, S. V. Ley and P. A. Almeida Paz. Facile, one-step production of niacin (vitamin B$_3$) and other nitrogen-containing pharmaceutical chemicals with a single-site heterogeneous catalyst, *Chem. Eur. J.*, **14**, 2340 (2008).

5. F. B. Mora and L. P. Llamedo. Experimental studies on the treatment of bone and joint tuberculosis with dihydrostreptomycin and isonicotinic acid hydrazine, *J. Bone Joint Surg. Am.*, **37**, 156 (1955).

6. R. Chuck. Technology development in nicotinate production, *App. Catal. A*, **189**, 191 (1999).

7. M. Hatanake and N. Tanaka (Nissan Chem. Ind. Ltd). WO 9305022 (1993).

8. R. Raja, J. M. Thomas, M. Xu, K. D. M. Harris, M. Greenhill-Hooper and K. Quill. Highly efficient, one-step conversion of cyclohexane to adipic acid using single-site heterogeneous catalysts, *Chem. Commun.*, 448 (2006).

9. L. F. Fieser and M. Fieser. *Organic Chemistry*, Reinhold, New York (1956).

10. J. Clayden, N. Greeves, S. Warren and P. Wothers. *Organic Chemistry*, OUP, Oxford (2001).

11. J. E. Backvall (ed.). *Modern Oxidation Methods*, Wiley-VCH, Weinheim (2004), and references therein.

12. F. Cavani, G. Centi, S. Perathoner and F. Trifiro (eds). *Sustainable Industrial Chemistry: Principles, Tools and Industrial Examples*, Wiley-VCH, Weinheim (2009).

13. W. Partenheimer. Methodology and scope of metal/bromide autoxidation of hydrocarbons, *Catal. Today*, **23**, 69 (1995).

14. L. Kesavan, R. Tiruvalam, M. H. A. Rahim, M. I. B. Saiman, D. I. Enache, R. L. Jenkins, N. Dimitratos, J. A. Lopez-Sanchez, S. H. Taylor, D. W. Knight, C. J. Kiely and G. J. Hutchings. Solvent-free

oxidation of primary carbon-hydrogen bonds in toluene using Au-Pd alloy nanoparticles, *Science*, **331**, 195 (2011).

15. R. Raja, J. M. Thomas and V. Dreyer. Benign oxidants and single-site solid catalysts for the solvent-free selective oxidation of toluene, *Catal. Lett.*, **110**, 179 (2006).

16. J. M. Thomas and R. Raja. Nanopore and nanoparticle catalysts, *Chem. Rec.*, **1**, 448 (2001).

17. J. M. Thomas and R. Raja. Catalytically active centres in porous oxides: design and performance of highly selective new catalysts, *Chem. Commun.*, 675 (2001).

18. R. Raja and P. Ratnasamy. Oxidation of cyclohexane over copper pthalocyanines encapsulated in zeolites, *Catal. Lett.*, **48**, 1 (1997).

19. R. Raja, J. M. Thomas, M. Xu, K. D. M. Harris, M. Greenhill-Hooper and K. Quill. Highly efficient one-step conversion of cyclohexane to adipic acid using single-site heterogeneous catalysts, *Chem. Commun.*, 448 (2006).

20. W. Niu, K. M. Draths and J. W. Frost. Benzene-free synthesis of adipic acid, *Biotechnol Prog.*, **18**, 201 (2002).

21. J. M. Thomas, R. Raja, B. F. G. Johnson, T. J. O'Connell, G. Sankar and T. Khimyak. Bimetallic nanocatalysts for the conversion of muconic acid to adipic acid, *Chem. Commun.*, 1126 (2003). (See also Editor's choice, *Science*, **300**, 867 (2003).)

22. R. Raja, G. Sankar and J. M. Thomas. Designing a molecular-sieve catalyst for the aerial oxidation of *n*-hexane directly to adipic acid, *Angew. Chem. Int. Ed.*, **39**, 2313 (2000).

23. R. J. Sheeman. Terephthalic acid dimethyl terephthalate and isophthalite acid, in *Ullmann's Encyclopedia of Industrial Chemistry*, online ed., Wiley-VCH, Weinheim (2005). http://onlinelibrary.wiley.com/doi/10.1002/14356007.a26_193.pub2/abstract

24. J. M. Thomas. Solid acid catalysts, *Sci. Am.*, **266**, 85 (UK ed.) (1992).

25. J. M. Thomas and D. W. Lewis. Towards rational design of solid acid catalysts, *Z. Phys. Chem.*, **197**, 37 (1996).

26. A. Takagaki, C. Tagusagawa, S. Hayashi, M. Hara and K. Domen. Nanosheets as highly active solid acid catalysts for green chemical syntheses, *Energy Environ. Sci.*, **3**, 82 (2010).

27. J. M. Thomas and R. Raja. Designing catalysts for clean technology, green chemistry and sustainable development, *Annu. Rev. Mater. Sci.*, **35**, 315 (2005).

28. F. Schüth. Engineered porous catalytic materials, *Annu. Rev. Mater. Sci.*, **35**, 209 (2005).

29. G. W. Huber, S. Iborra and A. Corma. Synthesis of transportation fuels from biomass; chemistry, catalysts and engineering, *Chem. Rev.*, **106**, 4044 (2006).

30. Y. Roman-Leshkov, J. N. Chheda and J. A. Dumesic. Phase modifiers promote efficient production of hydroxymethylfurfural from fructose, *Science*, **312**, 1933 (2006).

31. E. Taarning, S. Saravanamurugan, M. S. Holm, J. Xiong, R. M. West and C. H. Christensen. Zeolite-catalyzed isomerization of triose sugars, *Chem. Sus. Chem.*, **2**, 625 (2009).

32. M. S. Holm, S Saravanamurugan and E. Taarning. Conversion of sugars to lactic acid derivatives using heterogeneous zeotype catalysts, *Science*, **328**, 602 (2010).

33. M. Moliner, Y. Roman-Leshkov and M. E. Davis. Tin-containing zeolites are highly active catalysts for the isomerization of glucose in water, *PNAS*, **107**, 6154 (2010).

34. A. Corma, L. T. Nemeth, M. Renz and S. Valencia. Sn-zeolite beta as a heterogeneous chemoselective catalyst for Baeyer–Villiger oxidations, *Nature*, **412**, 423 (2001).

35. A. Corma. Attempts to fill the gap between enzymatic homogeneous and heterogeneous catalysis, *Catal. Rev.*, **46**, 369 (2004).

36. G. Strukul. Transition metal catalysts in the Baeyer–Villiger oxidation of ketones, *Angew. Chem. Int. Ed.*, **37**, 1198 (1988).

37. J. M. Thomas. Design, synthesis and *in situ* characterization of new solid acid catalysts, *Angew. Chem. Int. Ed.*, **38**, 3588 (1999).

38. R. Raja, J. M. Thomas and G. Sankar. Baeyer–Villiger oxidations with a difference: molecular sieve redox catalysts for the low-temperature conversion of ketones to lactones, *Chem. Commun.*, 525 (1999).

39. H. Kimura. Selective oxidation of glycerol on a platinum-bismuth catalyst by using a fixed bed reactor, *Appl. Catal. A.*, **105**, 147 (1993).

40. (a) C. H. Christensen, J. Rass-Hansen, C. C. Marsden, E. Taarning and K. Edeblad. Polymers from renewable sources: a challenge for the future, *Chem. Sus. Chem.*, **1**, 75 (2008).

    (b): P.N.R Vennestrom, C.M. Osmundsen, C.H. Christensen and E. Taarning. Beyond petrochemicals. The renewable chemicals industry, *Angew. Chem. Int. Ed.*, **50**, 10502 (2011).

41. Y. Roman-Leshkov, C. J. Barrett, Z.-Y. Liu and J. A. Dumesic. Mechanism of glucose isomerization using a solid Lewis acid, *Nature*, **447**, 982 (2007).

42. Y. Fan, C. Zhou and X. Zhu. Selective catalysis of lactic acid to produce commodity chemicals, *Catal. Rev.*, **51**, 293 (2009).

43. J. C. Serrano-Ruiz and J. A. Dumesic. Catalytic processing of lactic acid over Pt/Nb$_2$O$_5$, *Chem. Sus. Chem.*, **2**, 581 (2009).

44. A. Baeyer and V. Villiger. Einwirkung des Caro'schen Reagens auf Ketone, *Ber. Dtsch. Chem. Ges.*, **32**, 3625 (1899).

45. M. Hudlicky. *Oxidations in Organic Chemistry*, ACS, Washington (1990), p.186.

46. T. Mukaiyama. Oxygenation of olefins with molecular oxygen catalyzed by low valent metal complexes, in *The Activation of Dioxygen and Homogeneous Catalytic Oxidation,* ed. D. H. R. Barton, A. E. Martell and D. T. Sawyer, Plenum, New York (1993), p. 133.

47. R. Raja, G. Sankar and J. M. Thomas. New catalysts for the aerobic selective oxidation of hydrocarbons: Mn^{IV}- and Co^{IV}-containing molecular sieves for the epoxidation of alkenes, *Chem. Commun.*, 829 (1999).

48. G. W. Parshall and S. D. Ittel. *Homogeneous Catalysis,* 2nd ed., Wiley-Interscience, New York (1992), p. 252.

49. A. Corma. State of the art and future challenges of zeolites as catalysts, *J. Catal.*, **216**, 298 (2003).

50. P. Serna, M. Lopez-Haro, J. Calvino and A. Corma. Selective hydrogenation of nitrocyclohexane to cyclohexanone oxime with hydrogen on decorated Pt nanoparticles, *J. Catal.*, **263**, 328 (2009).

51. S. Saxena, J. Basak, N. Hardia, R. Dixit, S. Bhadauria, R. Dwivedi, R. Prasad, A. Soni, G. S. Okram and A. Gupta. Ammoximation of cyclohexanone over nanoporous TS-1 using UHP as an oxidant, *Chem. Eng. J.*, **132**, 61 (2007).

52. R. Raja, G. Sankar and J. M. Thomas. Bifunctional molecular sieve catalysts for the benign ammoximation of cyclohexanone: one-step, solvent-free production of oxime and ε-caprolactam with a mixture of air and ammonia, *J. Am. Chem. Soc.*, **123**, 8153 (2001).

53. J. M. Thomas and R. Raja. Design of a "green" one-step catalytic production of ε-caprolactam (precursor of nylon-6), *PNAS*, **102**, 13732 (2005).

54. R. Raja, J. M. Thomas and J. Paterson (unpublished work).

55. R. Mokaya and M. Poliakoff. The green production of nylon 6, *Nature*, **437**, 1243 (2005).

PART III

MESOPOROUS OPEN STRUCTURES

CHAPTER 6

EPOXIDATIONS AND SUSTAINABLE UTILIZATION OF RENEWABLE FEEDSTOCKS, PRODUCTION OF VITAMIN E INTERMEDIATES, CONVERSION OF ETHENE TO PROPENE AND SOLVENT-FREE, ONE-STEP SYNTHESIS OF ESTERS

6.1 Introduction

Already in earlier chapters we have seen that the catalytic performance of microporous (zeolitic) SSHCs can be significantly improved by pre-treating the nanoporous solid — e.g. by dealumination or steaming a well-defined aluminosilicate zeolite such as Y or mordenite — so as to introduce pores of larger diameter into the crystalline framework. What we shall focus upon in this chapter is a kind of nanoporous silica which, by adroit preparation, can be produced with pore diameters that may be controllably adjusted to fall (in a sharply defined range) anywhere within 20 to 500 Å. As mentioned in Chapter 1, a supreme advantage that SSHCs have over their homogeneous analogues is that, by suitable manipulation of the support, great scope exists for the engineering and characterization of desired active centres.

From the early days of heterogeneous catalysis, silica has been widely used as a support. Its merits are manifold: it is robust; it is readily prepared (with very high or low surface areas as required); it does not swell in contact with organic solvents; it has considerable thermal stability and attrition resistance as well as structural flexibility; and it bonds rather readily to a large number of the elements of the periodic table. All these qualities apply to the non-porous silicas (such as Aerosil and Cabosil) that have in the main been used in various studies of surface organometallic chemistry, as we describe in Chapter 7. When mesoporous silicas became routinely available[1,2] in the early 1990s, a radically new dimension was added to the use of silica supports, as described by Thomas[3] and Maschmeyer et al.,[4] in that

- ordered mesopores, epitomized by those present in MCM-41, MCM-48, KIT-*n*, MSU-*n* or SBA-15 (see Figure 1.3), that effectively convert the solid silica to one with a three-dimensional (3D) surface, and with a total area 5 to 50 times as large as the area of non-porous silicas, could be routinely prepared;
- the diameters of the mesopores can be controllably adjusted, as stated, and be prepared so that they do or do not intersect. (In both MCM-48 and SBA-15 the channels intersect.) All this assists diffusion of reactants into and products out of the sphere of reaction; and
- large organometallic and other (e.g. enzyme) molecules can be introduced into these pores to act either as catalysts in their own right (e.g. enantioselectivity as described in Chapter 7) or as precursors that serve the purpose of locating precisely anchored single atoms or anchored clusters as designed catalytically active (single-site) centres.

Above all, however, is the supremely convenient fact that, because mesoporous-silica-supported catalysts are ones in which all the active centres are distributed uniformly over a 3D surface within the catalyst body, one no longer has to restrict the techniques for catalyst characterization and *in situ* monitoring to surface-sensitive methods (like reflection IR or surface NMR), and *one has now the freedom to deploy all the traditional tools of solid-state chemistry and physics*, i.e. by resorting to X-ray powder diffraction and X-ray absorption, NMR, IR and UV transmission spectroscopies. This advantageous state of affairs that exists with open-structure solids — and this

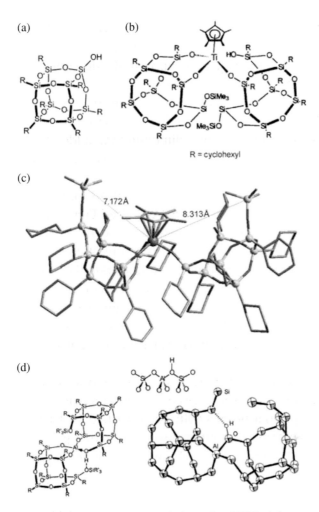

Figure 6.1 (a) Silsesquioxanes (general formula $(RSiO_{1.5})_n$) are structurally similar to β-cristobalite and β-tridymite (shown is the structure of $C_{y8}Si_8O_{12}$ (Cy = cyclohexyl). They are, effectively, soluble analogues of silica; and a heteroatom (usually transition metal or Ge) attached to the silsesquioxane simulates the behaviour (in solution) of the same, grafted atom attached to mesoporous silica forming the corresponding heterogeneous catalyst. (b) Fragment of the crystal structure (schematic) of a Ti^{IV}-cyclopentadienyl complex bound to two silsesquioxane moieties. (c) Actual structure (shown in (b)), indicating the isolated nature of the Ti^{IV} centre (hydrogen atoms have been omitted for clarity). (d) A soluble Brønsted acid site formed from silsesquioxanes.[29] (After Thomas *et al.*[42])

applies also to uniformly distributed active centres in microporous (zeolite) catalysts[5,6] (see Chapters 4 and 5) — is illustrated in Figure 1.3.

A further advantage in working with SSHCs that are located on a silica (internal and/or external) surface is that, thanks largely to the work of Feher *et al.*,[7,8] soluble analogues of silica, the so-called silsesquioxanes, are available (see Figure 6.1). These are, effectively, small fragments of crystalline silica comparable to tridymite and cristobalite. Consequently, by using the appropriate derivatized silsesquioxane (with its active centre attached to one of its vertices) one has a homogeneous analogue of the solid heterogeneous catalyst bearing an *identical* active centre, so that, for the first time, it becomes routinely possible to compare the catalytic performance of a homogeneous and heterogeneous catalyst where the active centres are the same (for example, see Figure 7.11). All this means that, with the abundance of various types of mesoporous silica, all having wall thicknesses separating the pores of *ca* 10 to 12 Å, and with all the internal surfaces of the walls bearing functionalizable, pendant Si-OH groups, as shown in Figure 6.2, it is readily possible to engineer the following distinct categories of SSHC[9]:

- single atoms or ions of a large number of elements;
- bimetallic clusters anchored to the internal area;
- immobilized asymmetric organometallic species; and
- hybrid nanoporous solids.

6.2 A Comprehensive Picture of the Nature and Mechanism of the Ti^{IV}-catalysed Epoxidation of Alkenes

In the early 1990s, my colleagues and I, along with Corma and his co-workers, showed how effectively spatially isolated Ti^{IV} sites, introduced during synthesis into nanoporous silica, were effective in epoxidizing cyclohexene.[4,10] This fact was not surprising in view of the Shell Company's earlier demonstration (in the patent literature) that a "Ti-SiO$_2$" catalyst was very effective for this and other conversions and also because of the known versatility (demonstrated

Mesoporous Silica

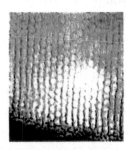 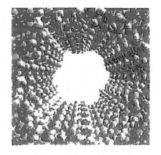

High-resolution electron micrograph of an ordered mesoporous silica, with pores of approximate diameter 100 Å and walls that are only a few atoms thick.	Each pore is lined with pendant silanol (Si–OH) groups. (Si, yellow; O, red; H, white)

Figure 6.2   Representation of the internal nature of a pore in mesoporous silica, replete with pendant silanol groups, SiOH.

by the EniChem Company) of TS-1 in selective oxidation of organic molecules (see Sections 4.5 and 4.6).

However, once it became technically possible in 1991 (in work carried out by the author and his team at the Davy Faraday Research Laboratory) to monitor both the short-range and long-range order of catalysts during, prior and after the acts of catalyst activation, catalytic turnover, catalyst poisoning and its reactivation — by recording XRD patterns and XAFS spectra in parallel[11] — altogether new possibilities were opened up in the characterization and design of SSHCs. In the first place one could test whether indeed the Ti^{IV} sites were spatially isolated. A significant advance was made when Ti^{IV} active centres were grafted onto the inner walls of mesoporous (MCM-41) silica using an organometallic precursor, in particular titanocene dichloride ($Ti(Cp)_2Cl_2$, where Cp stands for C_5H_5) as was done by Maschmeyer *et al.*[4] in 1995. The key steps in the introduction of the isolated, single-site active centres on the inner walls of the mesoporous silica are shown in Figure 6.3. The detailed course of this "heterogenization" of a Ti^{IV} active centre was followed by *in situ* X-ray absorption spectroscopy combined with *in situ* X-ray diffraction and FTIR. It is seen that a tripodally grafted Ti(OH) group constitutes the isolated active site (very resistant to leaching in this instance). Note that the van der Waals envelopes of the cyclopentadienyl groups attached to the surface-bound ("half-sandwich") intermediate, shown also in Figure 6.3 (top right), ensures that no two Ti[IV] active centres are closer than *ca* 7 Å, thereby ensuring the generation of a genuine, isolated, single-site catalyst. (The XAFS data leave no doubt that the titanyl group (>Ti=O) once proposed by Sheldon[12] as the active site for the epoxidation of alkenes is both not seen experimentally and computationally found to be energetically unfavourable.) This catalyst is an exceptionally good epoxidation one using alkyl hydroperoxides. Improvements can, however, be made to the performance of the Ti[IV]-centred epoxidation catalyst, and this can be achieved by arranging for a hydrophobic environment to be in the immediate vicinity of the Ti centre. Several ways of doing this are possible — see Oldroyd *et al.*[13] and Corma *et al.*[14,15]

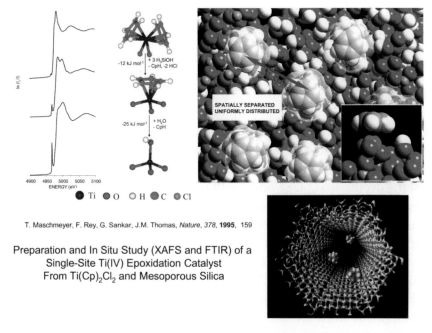

T. Maschmeyer, F. Rey, G. Sankar, J.M. Thomas, *Nature, 378,* **1995,** 159

Preparation and In Situ Study (XAFS and FTIR) of a
Single-Site Ti(IV) Epoxidation Catalyst
From Ti(Cp)$_2$Cl$_2$ and Mesoporous Silica

Figure 6.3　Summarized illustration of the preparation of spatially isolated titanol OH-Ti-(OSi)$_3$ active centres. (After Thomas *et al.*[9])

We shall return to this aspect of catalyst design later in the chapter, especially in the light of the remarkable advance made by Guidotti et al.[16] in their manner of rendering Ti^{IV}-centred catalysts (prepared from titanocene dichloride[4]) capable of effecting epoxidation of alkenes with H_2O_2 in an environmentally sustainable fashion.

6.2.1 *Mechanism of the Ti^{IV}-centred epoxidation of alkenes*

In situ XAFS studies[4,17,18] (see Figure 6.4) shed light on the mechanism of the epoxidation of cyclohexene with *t*-butyl hydroperoxide (abbreviated to HOOR[1] here). It transpires that this is an example of a Eley–Rideal process: the hydroperoxide is first attached to the active site, thereby converting the original four-coordinate Ti^{IV} to a six-coordinate one at steady state. The incoming alkene plucks one of the oxygens of bound hydroperoxide and liberates the alkene-epoxide while simultaneously restoring the "empty" (four-coordinate) active site. All the experimental and interpretive details are given in Figures 6.5, 6.6 and 6.7, and a more graphic picture is shown in Figure 6.6. The quantitative aspects of this mechanism are encapsulated in Figure 6.4, along with the essential XAFS data on which they are based.[18,19]

6.2.1.1 *The catalytic influence of replacing one of the Si atoms surrounding the Ti^{IV} active centre with Ge*

It is possible to boost the catalytic activity of the tripodally anchored titanol active centre $((\equiv SiO)_3 TiOH)$ by replacing one of the surrounding Si atoms with a Ge. We pre-treated the mesoporous silica with the appropriate amount of tetrabutyl germanium (Bu_4Ge) before introducing the titanocene dichloride ($TiCp_2Cl_2$) and then calcining.[20(a)] Moreover, because we know the precise atomic nature of the Ti^{IV}-centred active site in the heterogeneous catalyst, and because of the existence of silsesquioxane analogues (which are soluble), we can design a homogeneous catalyst that has the same active site as its heterogeneous counterpart.[20(b)] All this makes it possible to make a direct comparison of the catalytic performance of the tetrahedrally coordinated active site — see Figure 6.7 and Table 6.1.

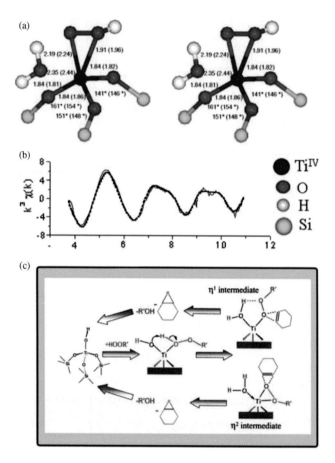

Figure 6.4 (a) XAFS analysis shows that, at steady state during the course of catalytic turnover, the coordination number of the Ti^{IV}-centred active site increases to six (all oxygens), with two of the oxygens at distances greater than 2.2 Å. (The distances and angles shown were derived from Ti K-edge EXAFS data, those in parentheses from DFT computation.) (b) Correspondence between experimentally determined EXAFS (full line) and computed data (dashed line) based on measurements made with the Ti^{IV}-centred catalyst during epoxidation. (c) On the basis of XAFS analysis and computations using DFT, this scheme is proposed for the epoxidation of an alkene by alkyl hydroperoxides (HOOR). Experimental evidence shows that both the η1 and η2 intermediates may be formed, and that the original (constrained) four-coordinate Ti active site passes through a six-coordinate state (e.g. the η2 intermediate) in the course of the epoxidation. The Ti^{4+} state remains as such during the "acid-base" process, wherein the oxygen of the hydroperoxide serves as the base.

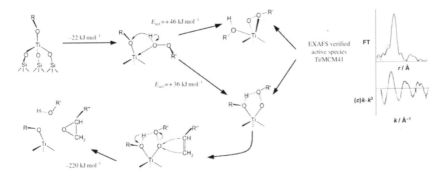

Figure 6.5 The catalytic cycle, including (DFT) calculated energies, for epoxidation of an alkene $R^1CH=CH_2$ to its epoxide for a four-coordinate Ti^{IV} active site at a mesoporous silica surface. (*In situ* XAFS measurements established the nature of the bare active sites and also the steady-state coordination of the Ti^{IV} centre.)

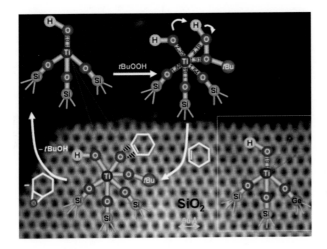

Figure 6.6 *In situ* EXAFS studies reveal the structure of the empty active site (a titanol group tripodally attached to mesoporous silica, top left) for the epoxidation of cyclohexene. The steady-state structure is sixfold coordinated (centre) from which the mechanism may be deduced. It is of the Eley–Rideal kind; only one of the two reactants (the *t*BuOOH) is adsorbed. The "free" cyclohexene plucks an oxygen atom from the hydroperoxide chemisorbed at the active site. When HOTi-$(OSi)_3$ active centres are replaced by HOTi-$(OSi)_2(OGe)$ (bottom right), a superior catalyst results.

Table 6.1 Comparison of the performance of insoluble heterogeneous single-site Ti^{IV}/SiO_2 epoxidation catalysts with their homogeneous soluble molecular analogues.

Homogeneous Catalysts	Heterogeneous Catalysts	TOF (h^{-1})
$[(c\text{-}C_5H_9)_7Si_7O_{12}Ti(OSiPh_3)]$		18
$[(c\text{-}C_5H_9)_7Si_7O_{12}Ti(OGePh_3)]$		52
	Ti↑SiO$_2$	26
	Ti↑MCM41	34
	Ti↑Ge↑MCM41	40

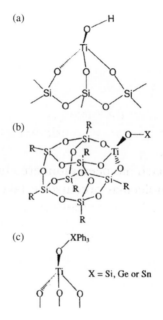

(a)

(b)

(c)   X = Si, Ge or Sn

Figure 6.7 Schematic representation of the Ti^{IV} centres (a) tripodally attached to silica, (b) within a titanosilsesquioxane ($R\text{-}Si_7O_{12}Ti\text{-}OX$), where R = cycloalkyl and X = Si, Ge or Sn, and (c) anchored onto silica as $(\geq Si\text{-}O)_3\text{-}Ti\text{-}OXPh_3$. (After Thomas *et al.*[20(b)])

6.2.1.2 *Comparing the catalytic activity of the same active site in a heterogeneous and a homogeneous catalyst*

The expoxidation kinetics, under identical experimental conditions, for the two heterogeneous catalysts anchored onto (a MCM-41) mesoporous silica, namely $(SiO)_3TiOH$ and $(SiO)_2(GeO)TiOH$, as well as for their soluble, homogeneous analogues, namely $RSi_7Ti(OSiPh_3)$ and $RSi_7O_{12}Ti(OGePh_3)$ (where R is the cyclopentyl group C_5H_9), are summarized in Table 6.1. Note that in the homogeneous catalysts, the active sites are $(SiO)_3TiOSi$ and $(SiO)_3TiOGe$. In reality the titanol group TiOH in the heterogeneous catalysts is replaced by TiOSi $\equiv$ and TiOGe $\equiv$.

It is seen that, within experimental error, the single-site activities in the epoxidation (of cyclohexene) are the same for the heterogeneous and homogeneous catalysts; and, moreover, when the tripodally bound Ti^{IV}-centred site is changed from $Ti-(OSi)_3$ to $Ti-(OSi)_2(OGe)$, the same trend in changed activity is seen for both the heterogeneous and homogeneous catalysts.

6.2.2 *An alternative method of introducing isolated Ti centres to mesoporous silica*

It must not be thought that the titanocene method is the only viable means of introducing single-site transition-metal (in this case Ti) active centres into mesoporous silica. An altogether different, but equally effective approach has been pioneered by Tilley, Bell *et al.*[21–24] in Berkeley, where a molecular precursor is taken to yield a series of active catalysts in silica. The metal ions in question are those of Ti, Cr, Fe and VO, and the essence of their preparation is that the desired atomic environment aimed at in the final catalyst (e.g. $Ti(OSi)_4$ or $Ti(OSi)_3$) is already present in the thermolytic precursor. Thus, by taking the tris(*tert*-butoxy) titanium complex (*i*Pro) $Ti[OSi(OtBu)_3]_3$, the environment ultimately achieved in the single-site catalyst is $Ti(OSi)_3$ and from $Ti[OSi(OtBu)_3]_4$ it is $Ti(OSi)_4$. Typical supports used by this group are MCM-41 and the more thermally stable SBA-15 mesoporous silicas. The general picture depicting the production of this type of SSHC is shown in Figure 6.8. In the case of an alkoxy (siloxy) species of the type $M[OSi(OtBu)_3]_n$, where M = Ti, Fe or Cr, this surface-attachment chemistry occurs with loss of HO*t*Bu or $HOSi(OtBu)_3$, to result in bonding to the surface through M-O-(surface) or Si-O-(surface) linkages,

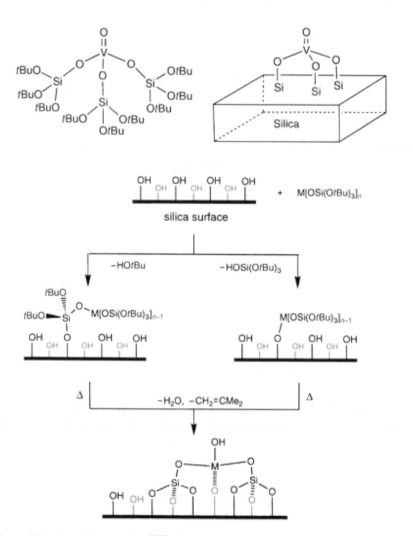

Figure 6.8 The Tilley method[22,23] of preparing single-site catalysts on mesoporous silica through thermolytic molecular precursors such as $M[OSi(OtBu)_3]$.

respectively. (Calcination results in the loss of all carbon and hydrogen, leading to the SSHC of nominal composition MO_x-$(n$-$1)SiO_2$.)

Che and Thomas *et al.*[25] introduced a Mo-centred SSHC onto mesoporous silica through the agency of molybdenocene (by analogy with titanocene). Other methods of creating transition-metal-centred SSHC on silica have been described by Dal Santo, Guidotti and co-workers.[26] And Tilley and his group[27] have extended his thermolytic molecular precursor (now abbreviated to TMP) to introduce Ta active centres, as well as others (onto mainly SBA-15 mesoporous silica). Tilley recognized that an "exposed" ion (such as Ti^{IV} or Ta^{V}) works best for epoxidation with non-aqueous oxidants (such as TBHP and cumene hydroperoxide (CHP)) because many of the crucial ion catalysts (exemplified by Ti^{IV}) are vulnerable to being poisoned by the coordination of water at the active site. Since H_2O_2 is a greener oxidant, one needs to devise a protection against poisoning of these ions in an aqueous H_2O_2 environment. For this reason Tilley has modified the active centre using trialkylsiloxy moieties. The resulting modified active centres were found to exhibit excellent selectivity for epoxide formation in the epoxidation of cyclohexene using aqueous H_2O_2 as the oxidant. It is to be noted that this epoxidation, too, follows Eley–Rideal kinetics. We shall see in the next section that Guidotti *et al.*[16] as well as Dutch workers van Santen and Abbenhuis,[28–30] using their so-called "hybrid" nanoporous SSHC, also succeeded in creating Ti^{IV} active centres capable of functioning satisfactorily in aqueous H_2O_2.

Much work has been done[31] to pin down more precisely the mechanism of the Ti^{IV}-centred epoxidation of alkenes by either alkyl hydroperoxide or H_2O as oxidants. The first key point, established experimentally by Marchese *et al.*,[32] in work subsequent to that carried out by Maschmeyer *et al.*,[4] is that individual Ti^{IV} ions, and not oligomers (involving $>Ti\langle^O_O\rangle Ti<$ linkages) or minute TiO_2 crystallites, are undoubtedly the active sites in the epoxidation. The second key point concerning mechanism, especially when the oxidant used in H_2O_2 and the titanol groups ($\equiv Si\text{-}O$)-TiOH is capped, as was done by Tilley and co-workers,[33(a)] is that the epoxidation occurs just as freely on the capped titanol centres (designated by them as $(SiO_{surf})_3Ti(OSiR_3)$) as on the uncapped ones. Tilley *et al.*[33a] used SBA-15 as their mesoporous silica. Their derived mechanism is shown in Scheme 6.1.

6.2.3 The use of H_2O_2 over Ti^{IV}-grafted mesoporous silica catalysts: a further step towards sustainable epoxidation[33]

Up until 2009, almost all the catalytic epoxidations conducted on the SSHCs in which Ti^{IV} centres were introduced by the titanocene

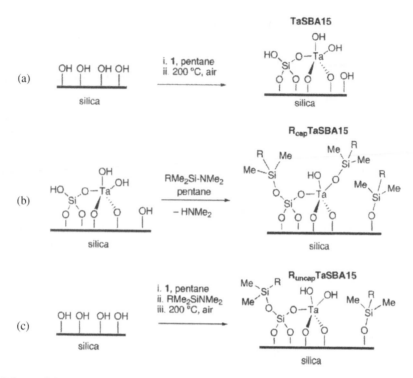

Scheme 6.1   Synthesis of TaSBA15, R$_{cap}$TaSBA15 and Ru$_{neap}$TaSBA15, where R ≡ Me, ^{n}Bu or ^{n}Oc. (After Ruddy and Tilley.[27])

method[4] (as outlined in Figure 6.3) were carried out using alkyl hydroperoxide (predominantly *t*-butyl or cumyl hydroperoxides) as oxidants. As we describe here, good catalytic performance was achieved with a range of alkenes, with alcoholic unsaturated terpenes[31,34,35] or unsaturated fatty acid methyl esters.[36,37] In the presence of H$_2$O$_2$, Ti^{IV} centres tend to suffer hydrolysis, quickly form aggregates, lose their activity and, to all intents and purposes, undergo irreversible deactivation.[38] The higher hydrophilic character and lower resistance to aqueous media of Ti^{IV}-containing mesoporous silicas, compared with the behaviour of TS-1 (microporous SSHCs) is clearly a drawback to the extensive use of the mesoporous Ti^{IV} SSHCs.

In a careful and important study by Guidotti *et al.*[33(b)] it was shown that, by the simple expedient of effecting the epoxidation (with H_2O_2) in a slow, dropwise fashion, so that there is a minimal local H_2O_2 concentration in the immediate vicinity of the Ti^{IV} active sites, a highly effective selective epoxidation takes place. They reported excellent selectivity (> 98%) in cyclohexene epoxide at the end of the slow addition of H_2O_2 in acetonitrite. Since cyclohexene, the alkene they studied, has been proposed by Noyori *et al.*[39] as a potential first step in the synthesis of adipic acid (where the alkene replaces cyclohexanone as the starting material), Guidotti *et al.* argued that this effective procedural expedient of theirs is progress towards sustainable epoxidation. Epoxides are, after all, extensively used as intermediates in all sectors of the chemical industry. Indeed, the epoxides that could, in future, be synthesized from naturally occurring fatty acid methyl esters (FAMEs), in a manner discussed in the next section, reinforces the important role Ti^{IV} centres grafted onto mesoporous silica could contribute materially to the quest for a sustainable chemical industry.

6.2.4 Ti^{IV} mesoporous catalysts have an important role to play in a sustainable way to utilize renewable feedstocks from fats and vegetable sources

The use of renewable feedstocks has been much publicized[40] in the context of green chemistry, clean technology and sustainability. Fatty acids and their derivatives are relatively underused as synthetic starting materials despite their natural abundance. Well over 1,000 natural fatty acids have now been identified; but the number that occur frequently and are present in commodity oils and fats is probably only about 25.[41] Moreover, the exploitation of fats from vegetable sources has many intrinsic advantages: not only are these sources eco-compatible, renewable and benign in their greenhouse gas effects, but, more important, they are convertible to several multifunctionalized molecules through a relatively small number of synthetic transformations. It has been apparent for some time (see, for example, Gunstone[41]) that the range of compounds obtainable from oils and fats may be extended, by varying the nature and/or position of the substitutents on the fatty acid hydrocarbon chain.

Epoxidized fatty acid derivatives from vegetable sources can be used in a variety of different ways. For example, they can be used as

- stabilizers and plasticizers in polymers;
- additives in lubricants; and
- components in plastics and in urethane foams, and as other intermediates for a variety of commodities.[37]

In a detailed study[37] of the FAME mixtures enumerated in Table 6.2, where only the most abundant oils are given, the four principal methyl esters that concern us are those shown in Scheme 6.2. All these esters could be readily epoxidized using the Ti^{IV} (MCM-41) catalyst consisting of single, spatially separated sites, prepared by the method of Maschmeyer *et al.*[4] using a titanocene precursor (see Figure 6.3).

The superior catalytic performance of the Ti^{IV}-MCM-41 silica (SSHCs) compared with commercial variants such as Ti-SiO_2

Table 6.2 Composition of the fatty acid methyl ester (FAME) mixtures.

		Composition (wt %)			
		HO[a] **Sunflower**	**Coriander**	**Caster**	**Soya-bean**
Palmitate	C16:0[b]	2	3	2	13
Stearate	C18:0	3	—	1	4
Oleate	C18:1	84	31	6	19
Linoleate	C18:2	10	13	3	56
Linolenate	C18:3	—	—	—	5
Bebenate	C22:0	1	—	—	1
Ricinoleate	C18:1-OH	—	—	87	—
Petroselinate	C18:1Δ^6	—	52	—	—
Others		—	1	1	2

[a] HO: high-oleic.

[b] Cxx:y, where xx is the number of carbon atoms of the fatty acid and y is the number of unsaturations. Δ^n, where n is the position of the unsaturation.

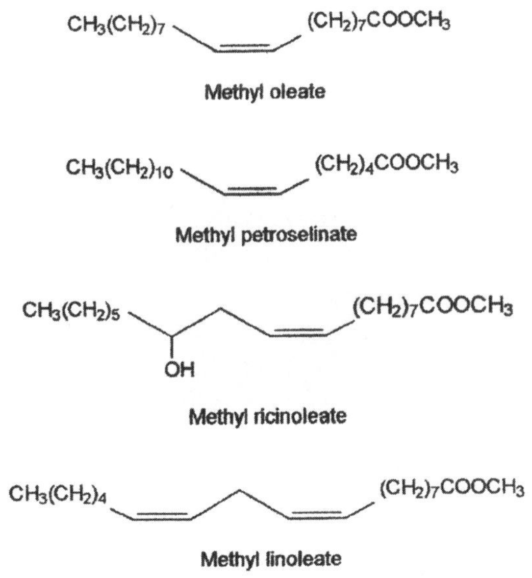

$$CH_3(CH_2)_7 \diagdown \diagup \diagdown (CH_2)_7COOCH_3$$

Methyl oleate

$$CH_3(CH_2)_{10} \diagdown \diagup \diagdown (CH_2)_4COOCH_3$$

Methyl petroselinate

$$CH_3(CH_2)_5 \diagdown \diagup \diagdown (CH_2)_7COOCH_3$$
$$\overset{|}{OH}$$

Methyl ricinoleate

$$CH_3(CH_2)_4 \diagdown \diagup \diagdown (CH_2)_7COOCH_3$$

Methyl linoleate

Scheme 6.2

(Davison Co.) or TiO_2-SiO_2 (Grace Co.), which have no ordered mesopores, is attributable[37] to the large amounts of highly accessible and well-defined Ti^{IV} single sites and the high density of silanols surrounding them. This accounts for the enhanced formation of epoxidized species when highly polar moieties (e.g. hydroxyl group in methyl ricinoleate) are present in the reacting feedstock.

6.3 Other Examples of Single-site, Metal-centred Catalysts Grafted onto Mesoporous Silica

Using the procedures developed by the author, by Tilley *et al.* and by others, a wide range of SSHCs involving mesoporous silica supports have been described. In the method of Tilley, as shown in Section 6.2.2, the desired atomic environment aimed at in the final catalyst (e.g. $Ti(OSi)_4$ or $Ti(OSi)_3$) is already present in the thermolytic precursor. Thus, for instance, starting from tris(*tert*-butoxy)siloxytitanium complex $(iPrO)Ti[OSi(OtBu)_3]_3$, the local environment achieved at the end in the single-site catalyst is $Ti(OSi)_3$ — see Figure 6.8.

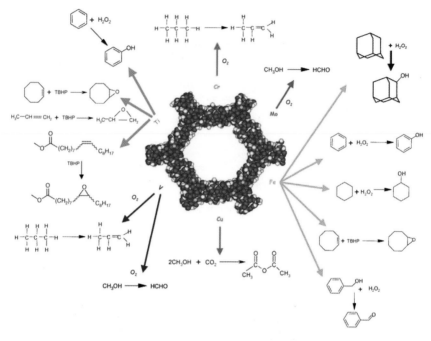

Figure 6.9 Illustration of some of the selective oxidations that may be effected by anchoring M^{n+} ions at the surfaces of mesoporous silica.[43]

By varying the catalytically active centre, it is possible to apply SSHCs in a large choice of selective oxidations.[26,27,42,43] These catalysts are of particular value in the pursuit of clean and sustainable chemistry.[26,42,44] Figure 6.9 shows a selection of SSHCs grafted onto the inner surface of mesoporous silica.

6.4 Titanium Cluster Sites for the Production of Vitamin E (Benzoquinone) Intermediates

Kholdeeva *et al.*[45,46] have shown that the vitamin E precursor, 2,3,5-trimethyl-1,4-benzoquinone (TMBQ), which is currently produced from 2,3,6-trimethylphenol (TMP), in an environmentally aggressive manner (using a $CuCl_2/O_2$ corrosive reagent that functions stoichiometrically), may be efficiently produced using aqueous H_2O_2 over a grafted Ti^{IV}-SiO_2 catalyst — see Scheme 6.3.

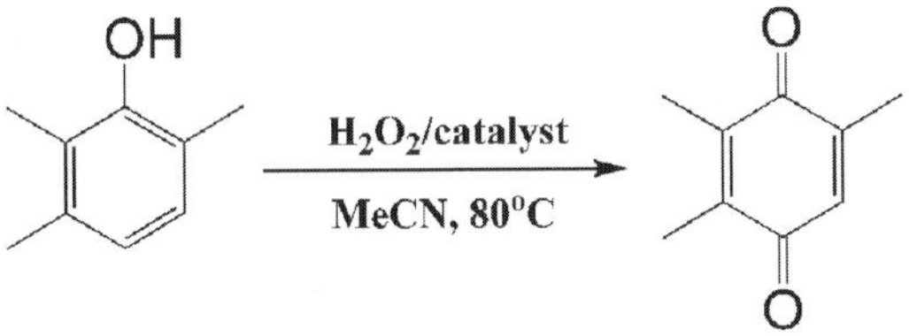

Scheme 6.3

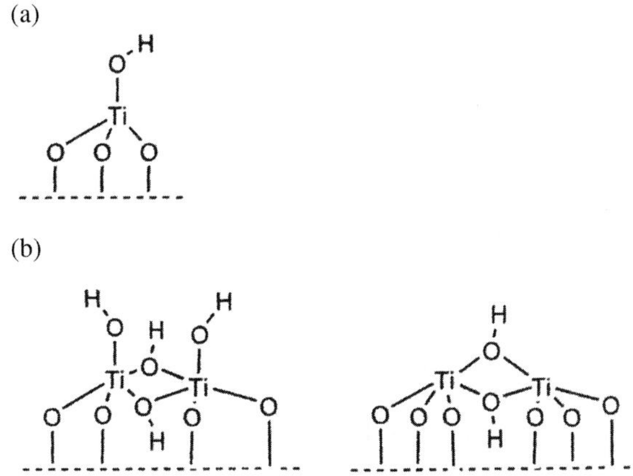

(a)

(b)

Scheme 6.4 (a) Isolated Ti single-site species and (b) possible dinuclear Ti species supported on silica surface. (After Kholdeeva *et al.*[45])

A detailed study by Kholdeeva, Guidotti and their co-workers,[47] who investigated the relative merits of mononuclear and dimeric Ti sites, came to the conclusion, as a result of DR-UV spectroscopic and other related studies, that the catalysts best capable of generating the desired benzoquinone possess Ti dimers distributed over the silica surface. (This study is also of relevance in several selective oxidations leading to the production of other fine chemicals, including the synthesis of vitamins E and Q, the co-enzyme Q and other valuable pharmaceutical products.)

The spectroscopic studies of Kholdeeva *et al.*,[47] in corroboration with those of Marchese *et al.*,[32] show that in their Ti[IV]-grafted catalysts there is an appreciable concentration of dinuclear Ti species, as shown in Scheme 6.4.

These workers observed that excellent selectivities (of 95% to 99%) in the conversion of TMP to TMBQ were obtained under conditions where there were high surface concentrations of the dinuclear Ti^{IV} species. Moreover, they found other evidence that favours a homolytic mechanism for the selective oxidation of the alkylphenol — as distinct from the heterolytic mechanism which holds for the epoxidation of alkenes at mononuclear Ti^{IV} sites (see Figures 6.4 and 6.5). All the facts garnered experimentally by Kholdeeva, Guidotti and co-workers led them to propose the mechanism shown in Scheme 6.5.

A dinuclear active site may undergo chemisorption of two H_2O_2 molecules and ensure a close proximity of two TiOOH species. With a low concentration of TMP and a high one for H_2O_2, it is plausible that one phenol molecule can be adsorbed by the reactive centre

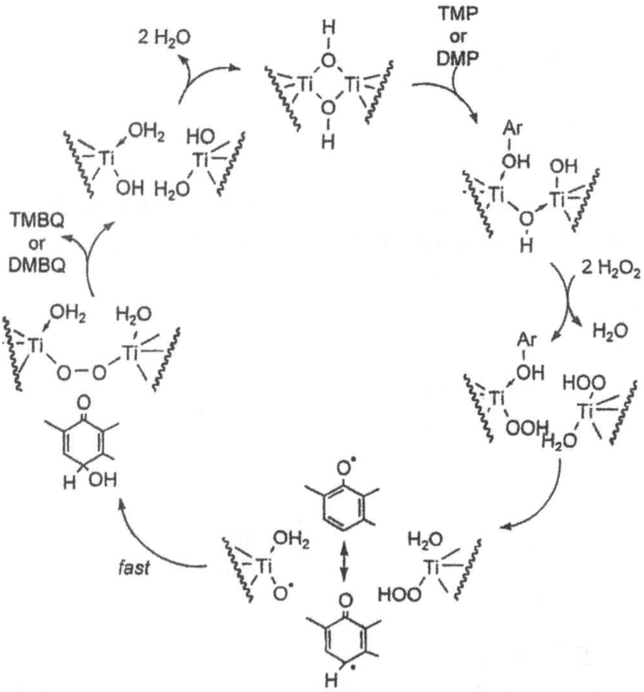

Scheme 6.5 Tentative mechanism of alkylphenol oxidation with H_2O_2 over a dimeric Ti site. (After Kholdeeva *et al.*[45])

that involves two adjacent hydroperoxo titanium groups. Interaction with the first TiOOH group produces a phenoxyl radical which is oxidized immediately by the second TiOOH group, giving rise to an intermediate quinol product, which, in turn, is oxidized rapidly to the final quinone. As a consequence, recombination of phenoxyl radicals leading to the undesired C-C and C-O dimeric by-products is minimized or suppressed over the dimeric Ti site, while the yield of the target quinone is enhanced compared to a monomer Ti site.

6.5 Single-site Metal Complexes Grafted onto Mesoporous Silica

Here we consider a few examples where well-known (as well as quite new) examples of catalytically (homogeneous) active metal compounds are anchored (grafted) to mesoporous silica surfaces. For heuristic purposes, we focus on three examples:

- MeReO$_3$ for epoxidation, studied by Herrmann, Kuhn and co-workers[48];
- a Co^{II} complex for selective oxidation of aryl aromatics studied by Bell, Tilley and co-workers[49]; and
- Ru carbenes for ring-closing and ring-opening metathesis reactions, studied by Balcar and co-workers.[50]

In their work involving methyltrioxorhenium (MeReO$_3$), in which the Re^{VII} species is anchored to mesoporous silica, Herrmann, Kuhn and co-workers found that the grafted catalysts were active in the epoxidation of alkenes using H$_2$O$_2$ as an oxidant, but that, during the course of repeated catalytic runs, some leaching associated with subsequent decomposition of the catalyst occurred.

Bell and Tilley's pseudo-tetrahedral Co^{II} complex was grafted onto mesoporous (SBA-15) silica as outlined in Scheme 6.6.

This catalyst for oxidations of ethylbenzene and other alkylaromatics using *tert*-butyl hydroperoxide as oxidant at 80°C exhibits good conversions to acetophenone. In the case of *sec*-phenyl alcohol it gives 100% conversion (and 100% selectivity) to acetophenone. The process is a free radical one in which two different initiation

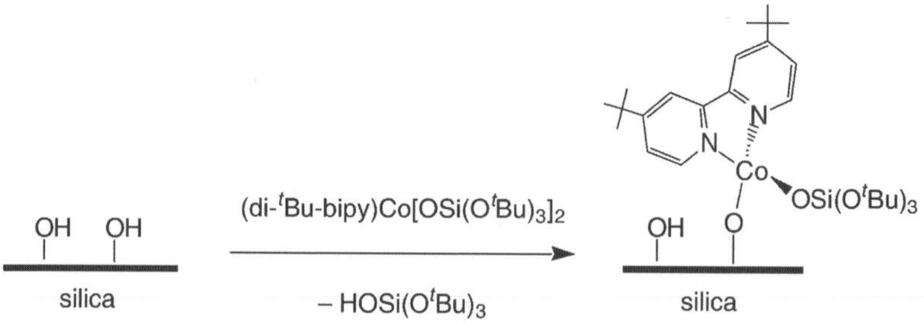

Scheme 6.6

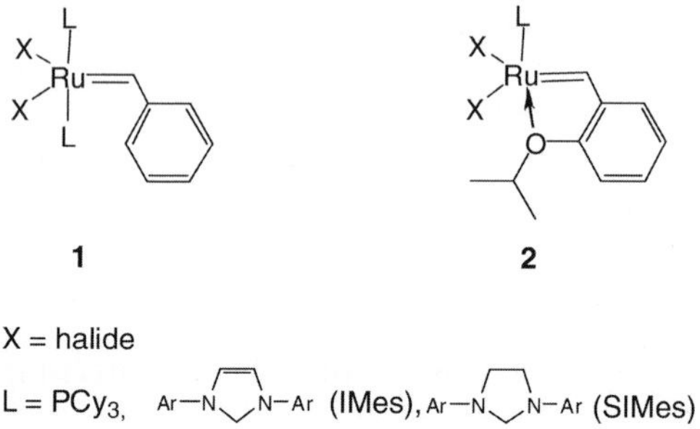

1   **2**

X = halide

L = PCy$_3$,   Ar—N⁀N—Ar (IMes), Ar—N⁀N—Ar (SIMes)

Scheme 6.7   Ru metathesis catalysts. (After Bek *et al.*[50])

steps are envisaged for the benzyl radical formation: (i) an electron transfer from the arene to Co^{III} to yield an arene radical cation, which then produces a benzyl radical upon H$^+$ loss, or (ii) H-atom transfer through abstraction of benzylic hydrogen atoms by radical species. Kinetic isotope studies favoured the H-atom mechanism.

In the studies of Balcar and co-workers, new hybrid catalysts for olefin metatheses were prepared by immobilization on the mesoporous silica of the Hoveyda–Grubbs-type catalysts shown in Scheme 6.7.

These catalysts exhibited high activity in ring-closing metathesis of 1,7-octadiene and diethyl diallylmalonate. They were reusable, and showed little tendency to leach. In the ring-opening metathesis polymerization of norbornene and cyclooctene (on SBA-15), the single-site catalyst gave higher yields of high-molecular-weight polymers than the corresponding homogeneous catalyst and also higher than when grafted to conventional silica supports. The reactions studied[50] are shown in Scheme 6.8.

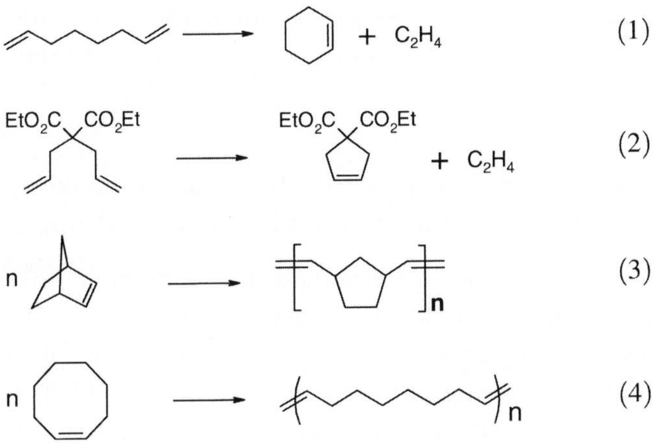

$$\text{(1)}$$

$$\text{(2)}$$

$$\text{(3)}$$

$$\text{(4)}$$

Scheme 6.8

6.5.1 *Stability and recyclability of supported metal-ligand complex catalysts: a critical note*

Although this section deals almost exclusively with non-porous silica (and other kinds of non-mesoporous supports) the following note of caution, sounded eloquently by Jones,[51] is also apposite here. It is a self-evident fact that heterogenized metal complex catalysts continue to offer tremendous potential for easing catalyst recovery while also allowing tunable, molecular-catalyst-based activity and selectivity. For 40 years reams of reports on this topic have been focused largely on demonstration-of-principle studies; but precious few have scrutinized the stability (longevity) and other incidental related facts (like susceptibility to catalyst poisoning) pertaining to the viability of a new catalyst. Greater emphasis in future studies needs to be placed on catalyst stability, robustness and attrition resistance before a new anchored catalyst becomes a viable proposition in a commercial sense. The kind of study, described by Jones, of the stability and proneness to deactivation of supported (enantioselective) Co-salen epoxide ring-opening catalysts needs to be undertaken also for the types of catalysts described earlier in this section (derived from mesoporous silica).

This cautionary note applies particularly to enantioselective catalysts, the subject of Chapter 7, consisting of chiral metal-ligand complexes.

6.6 A Trifunctional, Mesoporous Silica-based Catalyst: Highly Selective Conversion of Ethene to Propene

As mentioned earlier in this monograph (see Section 4.6), the idea of multifunctionality is closely related to the practice of effecting consecutive (i.e. a cascade of) reactions in "one pot". Reactions of this kind are of growing importance; and it is very likely that, with mesoporous silica as the framework, and the ability to graft[26–28,44,52,53] onto the inner surfaces isolated metal ions and oxy ions (and small metal clusters — as we shall describe in Chapter 8), there is abundant scope[54] to construct many new mono-, bi- and multifunctional single sites.

A remarkable trifunctional catalyst has recently been constructed by Iwamoto and Kosugi.[55] This catalyst is capable of executing dimerization, isomerization and metathesis, all occurring in the nanospace within a mesoporous silica (see Figure 6.10). To my

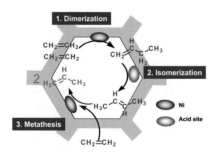

- Reaction of ethylene on Ni-MCM41
  - At 573 K to butenes, conv. 43%, select. 93%
  - At 673-723 K to propylene, conv. 53%, select. 55%

Figure 6.10 Iwamoto and Kosugi[55] have developed a trifunctional mesoporous siliceous catalyst where dimerization (of ethene), followed by isomerization and then metathesis occur to form propene.

knowledge, there is no homogeneous catalyst known that can effect this cascade of three reactions to achieve the desirable conversion of (plentifully available) ethene to industrially highly desirable propene. Here, unlike the situation that prevails in the case of TAPO-5 catalysed conversion of cyclohexene to adipic acid (see Section 4.6), where only one multifunctional active site exists, there are believed to be two distinct sites within a single nanoreactor: an acid site located at the inner wall, where an Al^{III} ion substitutes for Si^{IV}, and a metal ion anchored to the wall, and thought to be[56] Ni^{III}. Iwamoto *et al.* report a 68% conversion in a single pass over the catalyst at 673 K, with a propene selectivity of 48%.

6.7 Hybrid SSHCs are Chemically Robust

Insofar as the stability, recyclability and general robustness of catalysts are concerned — the points emphasized in Chapter 1 and Section 6.5.1 — there is a great deal to be said in favour of the kind of hybrid catalysts developed by Abbenhuis and van Santen.[28–30] It will be recalled that the Ti^{IV}-centred, tripodally anchored epoxidation catalysts (see Figure 6.4) function at their best in a hydrophobic environment with oxidants such as TBHP and CUHP. Even though Guidotti *et al.*[33(b)] (see Section 6.2.3) found a reliable method — like adding the reagents dropwise and slowly — that could use this catalyst with aqueous H$_2$O$_2$ as an oxidant, it is better to modify the immediate environment of the Ti^{IV} active site with "protective" groups that will render them more robust in coping with aqueous H$_2$O$_2$ as the oxidant. This is the tactic also used by Tilley and his co-workers. When the Ti^{IV} active centre is in a hydrophobic cavity it is not poisoned by water, so, by first incarcerating a Ti-silsesquioxane inside a mesoporous silica (Figure 6.11) and finding that it gave rise to an even more robust catalyst, Abbenhuis and van Santen proceeded to design (in LEGO fashion!) nanoporous siliceous solids that were lined with (hydrophobic) organic films, following recipes evolved by Ryoo *et al.*,[57] which were then chemically bonded to a siloxane framework as schematized in Figures 6.12 and 6.13.

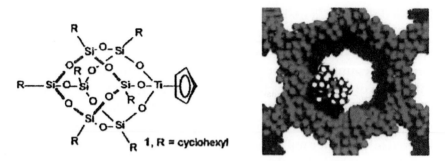

Figure 6.11 A more active single-site Ti[IV]-centred catalyst was produced by Abbenhuis and van Santen[28] by adsorbing a silsesquioxane-Ti[IV] complex on the inner walls of mesoporous silica.

To understand fully what has been accomplished here, we first recall some salient properties of silsesquioxanes (see Figure 6.1). Thanks to the work of Maschmeyer *et al.*, involving high throughput testing, it is now relatively easy to prepare silsesquioxanes with a wide range of functional groups attached to their framework. Moreover, an enormous number of single metallic (or non-metallic) elements can be readily incorporated into the framework of these so-called polyhedral oligomeric silsesquioxanes (POSS), thanks largely to the pioneering work of the Eindhoven group. Figure 6.14 summarizes the wealth of such substitutions that may be wrought in the large family of silsesquioxane solids. Note, too, that by taking partially complete POSS entities, one has control over how many adjacent pendant silanol groups can be "chosen" for subsequent functionalization. Another important point to make is that the steric and electronic properties of the silsesquioxane ligands render metal centres more Lewis acidic than conventional alkoxides or siloxide ligands do.

6.8 The Confluence of Heterogeneous and Homogeneous Catalysis Involving Single Sites

Figure 6.11 depicts the way in which, using Ryoo's recipe, Abbenhuis and van Santen were able to insert a hydrophobic (organic) lining inside the mesoporous silica, MCM-41.

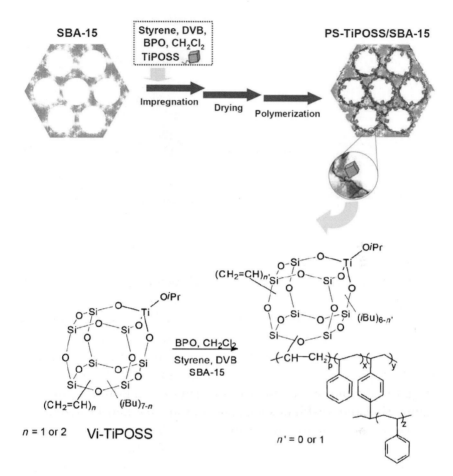

Figure 6.12   Schematic outline of the synthesis of a three-dimensionally netted polymeric (hybrid) catalyst formed in the procedure evolved by the Eindhoven–Dalian collaboration.[28]

In Figure 6.12, the 3D nanoporous solid SBA-15 was used as the "accessible" diffusive pathway for reactants to reach the hybrid catalysts, formed by anchoring the functionalized POSS units onto the inner walls of the modified silica. The convergence of homogeneous and heterogeneous catalysis, illustrated in Figures 6.13 and 6.14, shows how it becomes possible to utilize the same aggregate of atoms that, in the functionalized silsesquioxane, constitutes the

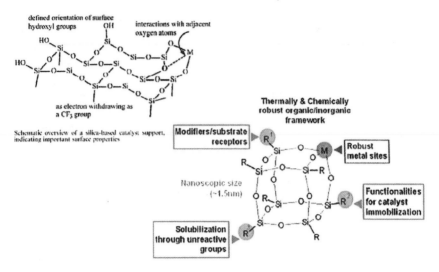

Figure 6.13 A summary of how a catalyst, formed by the convergence of homogeneous and heterogeneous catalysis, functions — see Abbenhuis and van Santen[28] for further details.

active site either as a homogeneous or as a heterogeneous catalyst. The hybrid catalyst is, in essence, one in which the nanoporous siliceous host (suitably pre-conditioned to form a hydrophobic pocket) houses the molecular catalyst as an integral part of the overall structure. Figure 6.14 summarizes how this approach of combining the best of inorganic SSHCs with aspects of homogeneous catalysis has led to commercial application.

A particularly elegant and robust hybrid SSHC, where the Ti^{IV} four-coordinate active centres are especially well protected and less vulnerable to leaching during use was devised by Marchese and co-workers. They prepared Ti-POSS-based heterogeneous catalysts by anchoring a functionalized Ti-containing silsesquioxane onto the inner surface of an ordered (as well as of a non-ordered) mesoporous silica.[59] A novel type of bifunctional POSS, containing both a Ti^{IV} metal centre and a triethoxysilane group for grafting (onto the silanol-rich surfaces of the mesopores), was synthesized by these workers and named Ti-POSS-TSIPI (see Scheme 6.9). The catalytic performance in epoxidations of the terpene limonene on this

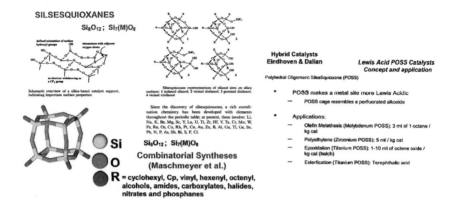

Figure 6.14 Hybrid single-site heterogeneous catalysts derived from silsequioxanes.

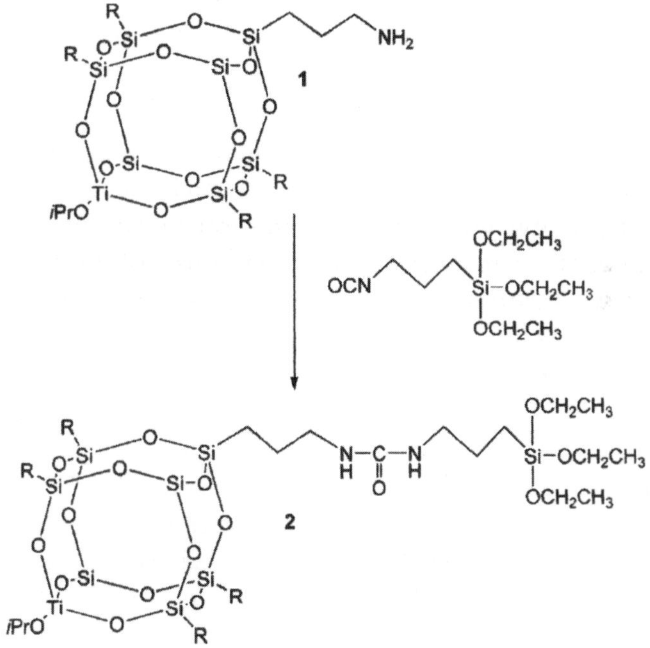

Scheme 6.9 The reaction between Ti-NH$_2$-POSS (**1**) and 3-isocyanatopropyl triethoxysilane for the preparation of Ti-POSS-TSIPI (**2**); R = isobutyl group. (After Carniato *et al.*[59])

anchored catalyst on both SBA-15 and a commercial non-ordered silica (e.g. Davison Co. mesoporous silica) was comparable to that of the Ti-MCM-41 preparation first reported by Thomas *et al.*,[9] Maschmeyer *et al.*,[4] Guidotti *et al.*[35] and others.

6.9 Beyond Mesoporous Silica

Whereas the preceding sections of this chapter have been restricted to the performance of catalysts derived from functionalizing mesoporous silica, there is growing interest now in mesoporous forms of oxides of metals like Nb, Tc, W (as predicted in an early paper[3]); and already a substantial body of interesting catalytic phenomena[60] exists using such nanoporous solids. Moreover, there is an equally impressive range of inorganic silica networks with organic functions — the so-called periodic mesoporous organosilicas (PMOs), described recently by Hoffmann and Fröba.[61] In addition, relevant single-site catalytic studies involving other mesoporous as well as mesocellular silica foams (MCFs), with pore diameters extending up to 500 Å, are already the subject of catalytic investigation.[53,62,63] And a new trend, initiated by Maschmeyer and co-workers,[64] entails designing hierarchically structured composite material with interconnecting meso- and micropores. This allows for various types of nanosized (microporous) zeotypes to be incorporated into a 3D mesoporous matrix. The particular proof of principle described by Maschmeyer *et al.* is a system in which zeolite Beta shows a significantly higher cracking activity per gram of zeolite than pure zeolite Beta for *n*-hexane. One must also recognize the great potential possessed by the ever-expanding family of metal-organic frameworks (MOFs) in the scope that exists for post-synthesis modification of the framework so as to "place" strategically useful (single-site) active centres in such open-structure solids.

6.9.1 *The merits of clay-based single-site catalysts*

It has long been known that sheet silicates — to be precise, sheet aluminosilicates — are capable of catalysing a wide range of organic reactions.[65-70] Naturally occurring clays like montmorillonite (idealized formula $M_x^+(Si_8)^{tet}(Al_{4-y}Mg_y)^{oct}O_{20}(OH)_4$) and beidellite (idealized formula $M_y^+(Si_{8-x}Al_x)^{tet}(Al_4)^{oct}O_{20}(OH)_4$) — which tend to contain many impurities — and their highly pure synthetic analogues exhibit a remarkable range of catalytic behaviour. Because the exchangeable, interlamellar ions M^+ may be readily replaced by highly polarizing ones, like Al^{3+} or La^{3+}, this results in the interlamellar region (the space separating the individual sheets) being the locus of cation hydrolysis (e.g. $M^{3+}(H_2O) \rightarrow M(OH)_2^{2+} + H^+$). This generates high Brønsted acid activity in the solid, the single sites being like La^{3+} in zeolite Y. The following are some of the industrially important reactions that may be effected by acid clay catalysts — see Scheme 6.10.

Clay catalysts of this kind are a convenient means of synthesizing a large range of alkanols, ethers, esters, amines and alkylated aromatics.[71]

The second of the reactions in Scheme 6.10 is the basis of the large-scale industrial production (in excess of 250×10^3 tonnes per annum) in the UK of an important commodity chemical, ethyl acetate.[72] (Note that this solvent-free, one-step, 100% atom-efficient catalysed reaction dispenses with the traditional method of ester production involving addition of alcohol to acid with water formed as by-product, and which requires an extra step to separate the product from the water.)

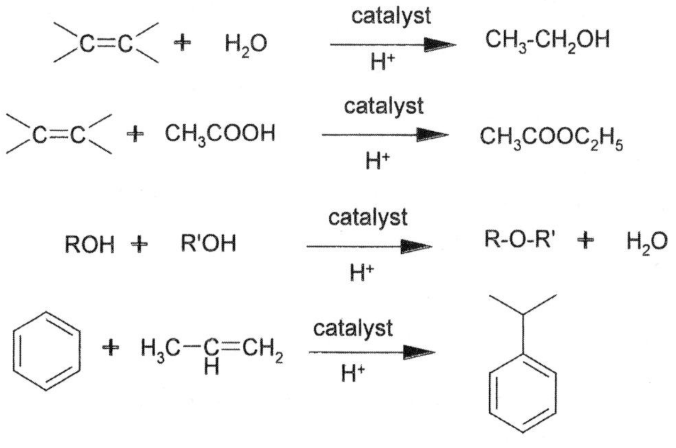

Scheme 6.10

Diddams *et al.*[68] showed how beidellite-montmorillonite layered silicates and their pillared analogues exhibited good catalytic activity in secondary amine formation from cyclohexylamine, in ester production from hex-1-ene and acetic acid, and in ether synthesis from pentanol — see Scheme 6.11 and Table 6.3.

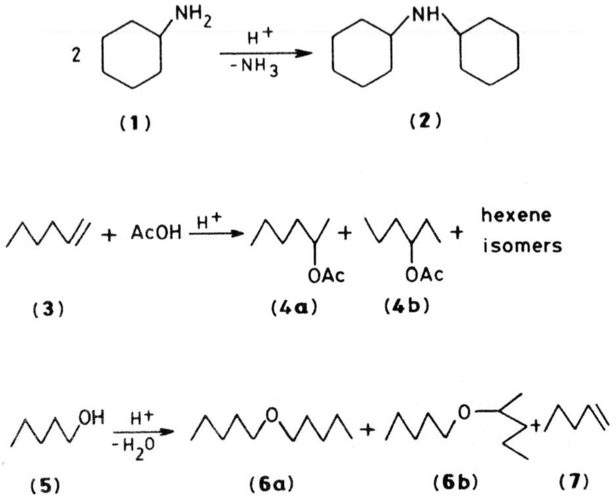

Scheme 6.11

Table 6.3 Catalytic performance of synthetic beidellite (B) and of naturally occurring montmorillonite (M) and their pillared analogues (all Al^{3+}-exchanged). Total yields expressed as mol %.

	B	M	Pillared B	Pillared M
Amine formation	24.9	27.9	6.5	1.7
(1) → (2)				
Hex-1-ene (3) + AeOH	12.0	34.1	2.2	0.3
Percentage of:				
hex-2-yl acetate (4a)	50.6	23.2	66.2	100
hex-3-yl acetate (4b)	14.2	10.2	9.4	—
hexene isomers	35.2	66.6	24.4	—
Ether formation from	42.0	54.5	6.3	1.9
n-pentanol (5)				
Percentage of:				
1,1-dipentyl ether (6a)	56.8	46.5	60.5	63.9
1,2-dipentyl ether (6b)	4.8	5.1	3.8	28.0
pent-1-ene (7)	38.4	48.4	35.7	8.1

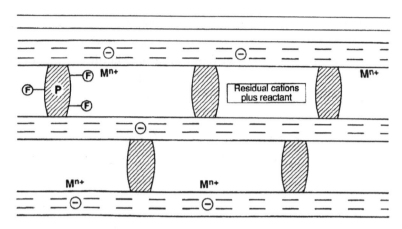

Figure 6.15 Layered aluminosilicates, when pillared, using bulky, multivalent oxo-inorganic cations, can be rendered effective high-area catalysts — see Thomas,[67] Diddams *et al.*,[68] Jones,[69] Ballantine *et al.*[72] and text.

Figure 6.15 shows how three-dimensional porosity is introduced into sheet-aluminosilicate catalysts by pillaring them. The nature of the pillar can vary quite widely: two popular (large) cations suitable for pillaring clays are $[(Al_{13}O_4(OH)_{24})(H_2O)_{12}]^{7+}$ and $[Zr_4(OH)_8(H_2O)_{16}]^{8+}$. After introducing these large cations into the interlamellar regions, the solid is calcined, thereby yielding alumina and zirconia pillars, respectively. Often these pillars are functionalized by subsequent appropriate treatment so as to increase the number of single sites in the resulting catalyst.

6.9.2 *Pillared zeolites?*

Recent reviews by Roth and Čejka[73(a)] and Tsapatsis *et al.*[73(b)] have drawn attention to the prospect of pillaring delaminated zeolites, such as that known as MCM-22, which is synthesized via a layered precursor. Whilst no specific catalyst applications have emerged from their formulated possibilities, it is nevertheless instructive to note the ease with which a 2D zeolite (MCM-22) can, by pillaring, be converted into a more open 3D network, as illustrated in Figure 6.16.

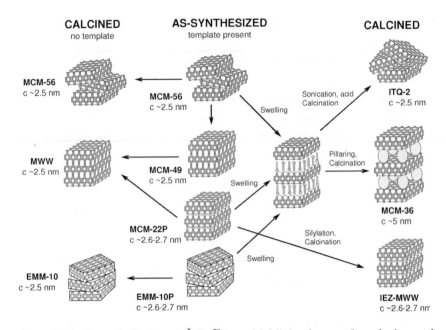

CALCINED
no template

AS-SYNTHESIZED
template present

CALCINED

MCM-56
c ~2.5 nm

MCM-56
c ~2.5 nm

Swelling

Sonication, acid
Calcination

ITQ-2
c ~2.5 nm

MWW
c ~2.5 nm

MCM-49
c ~2.5 nm

Swelling

Pillaring,
Calcination

MCM-36
c ~5 nm

MCM-22P
c ~2.6-2.7 nm

Silylation,
Calcination

Swelling

EMM-10
c ~2.5 nm

EMM-10P
c ~2.6-2.7 nm

IEZ-MWW
c ~2.6-2.7 nm

Figure 6.16 Recently Roth and Čejka[73] have highlighted ways of producing and elaborated the advantages of pillared zeolitic material like the MCM-36 product, with a (permanent) interlamellar separation of *ca* 50 Å.

It has been reported that some ten zeolitic frameworks have the potential of being pillared in the manner described by Roth and Čejka. Useful scope exists here for the design of new single-site heterogeneous catalysts, just as with pillared clays.

6.10 Envoi

The great advantages that mesoporous silicas possess have been exemplified repeatedly in this chapter. Not only can SSHCs designed using mesoporous silica cope with bulkier reactants than those catalysed by microporous SSHCs, they also allow bulky precursor molecules (typified by organometallic species such as titanocene or molybdenocene salts) to be utilized so as to form the spatially isolated single sites in the first instance.

Ti^{IV}-centred SSHCs have figured conspicuously in our discussion, partly because of the importance of such catalysts in producing key chemical intermediates such as epoxides; partly also because Ti^{IV}-centred mesoporous catalysts have a demonstrated[33] key role to play in utilizing feedstocks such as fats and vegetable oils for a sustainable chemical industry, and for the production of certain vitamins.[47] Ti^{IV}-centred mesoporous silica work best as oxidation catalysts with alkyl hydroperoxide oxidants, but good progress has been made by Guidotti *et al.* and van Santen *et al.* to make them more robust for use with aqueous H_2O_2 oxidants, a step in the right direction so far as the trend for sustainability and environmental responsibility is concerned.

Another major reason for focusing on Ti^{IV}-centred SSHCs is that they are amenable to detailed quantum chemical analysis of the energetics of their formation and mode of action.[74] Such calculations, aided by kinetic and spectroscopic studies, allow us to arrive at a plausible mechanism of epoxidation of alkenes (see, e.g. Figure 6.5).

But mesoporous silica offers several opportunities for designing other single-site catalysts, centred on many other metals (Ta, Cu, Mo, Mn, Vo, Fe, Ni, etc., see Figure 6.9). They also allow robust new hybrid catalysts to be designed, as described in the work of van Santen, Li and co-workers,[30] and hierarchical ones in which nano-particles of a zeolite are accommodated within a mesoporous host.[64] In addition, as shown by Iwamoto and Kosugi,[55] a trifunctional catalyst may be designed using mesoporous silica that converts ethene in high yield to propene.

The catalysts designed here using mesoporous silica also remind us that the distinction between heterogeneous and homogeneous catalysts is often blurred.[75–78]

References

1. C. T. Kresge, M. E. Leonowicz, W. J. Roth, J. C. Vartuli and J. S. Beck. Ordered mesoporous molecular sieves synthesized by a liquid-crystal template mechanism, *Nature*, **359**, 710 (1992).
2. D. Zhao, Q. Huo, J. Feng, B. F. Chmelka and G. P. Stucky. Nonionic triblock and star diblock copolymer and oligomeric surfactant syntheses

of highly ordered, hydrothermally stable, mesoporous silica structures, *J. Am. Chem. Soc.*, **120**, 6024 (1998).

3. J. M. Thomas. The chemistry of crystalline sponges, *Nature*, **368**, 289 (1994).

4. T. Maschmeyer, F. Rey, G. Sankar and J. M. Thomas. Heterogeneous catalysts obtained by grafting metallocene complexes onto mesoporous silica, *Nature*, **378**, 159 (1995).

5. J. M. Thomas. Uniform heterogeneous catalysts: the role of solid-state chemistry in their development and design, *Angew. Chem. Int. Ed.*, **27**, 1673 (1988).

6. J. M. Thomas. The Bakerian Lecture, 1990: new microcrystalline catalysts, *Phil. Trans. R. Soc.*, **333**, 173 (1990).

7. F. J. Feher and R. L. Blanski. Olefin polymerization by vanadium-containing silsesquioxanes: synthesis of a dialkyl-oxo-vanadium(V) complex that initiates ethylene polymerization, *J. Am. Chem. Soc.*, **114**, 5886 (1992).

8. F. J. Feher and T. L. Tajima. Synthesis of a molybdenum-containing silsesquioxane which rapidly catalyzes the metatheses of olefins, *J. Am. Chem. Soc.*, **116**, 2145 (1994).

9. J. M. Thomas, R. Raja and D. W. Lewis. Single-site heterogeneous catalysts, *Angew. Chem. Int. Ed.*, **44**, 6456 (2005).

10. G. Sankar, F. Rey, J. M. Thomas, G. N. Greaves, A. Corma, B. R. Dobson and A. J. Dent. Probing active sites in solid catalysts for the liquid-phase epoxidation of alkenes, *Chem. Commun.*, 2279 (1994).

11. J. W. Couves, J. M. Thomas, D. Waller, R. H. Jones, A. J. Dent, G. E. Derbyshire and G. N. Greaves. Tracing the conversion of aurichalcite to a copper catalyst by combined X-ray absorption and diffraction, *Nature*, **354**, 465 (1991).

12. R. A. Sheldon. New catalytic methods for selective oxidation, *J. Mol. Catal.*, **20**, 1 (1983).

13. R. D. Oldroyd, J. M. Thomas and G. Sankar. Engineering an atomically well-defined active site for the catalytic epoxidation of alkenes, *Chem. Commun.*, 2025 (1997).

14. A. Corma, M. T. Navarro and F. Rey. One-step synthesis of highly active and selective epoxidation catalysts formed by organic-inorganic Ti-containing mesoporous composites, *Chem. Commun.*, 1899 (1998).

15. A. Corma, M. Domine, J. A. Gaona, J. L. Jorda, M. T. Navarro, F. Rey, J. Perez-Pariente, J. Tsuji, B. McCulloch and L. T. Nemeth. Strategies to improve the epoxidation activity and selectivity of Ti-MCM-41, *Chem. Commun.*, 2211 (1998).

16. M. Guidotti, G. Pirovano, N. Ravasio, B. Lazaro, J. M. Fraile, J. A. Mayoral, B. Coq and A. Galarneau. The use of H_2O_2 over titanium-grafted mesoporous silica catalysts: a step further towards sustainable epoxidation, *Green Chem.*, **11**, 1421 (2009).

17. J. M. Thomas and G. Sankar. The role of synchrotron-based studies in the elucidation and design of active sites in titanium-silica epoxidation catalysts, *Acc. Chem. Res.*, **34**, 571 (2001).

18. G. Sankar, J. M. Thomas, C. R. A. Catlow, C. M. Barker, D. Gleeson and N. Kaltsoyannis. The three-dimensional structure of the titanium-centered active site during steady-state catalytic epoxidation of alkenes, *J. Phys. Chem. B.*, **105**, 9028 (2001).

19. J. M. Thomas, C. R. A. Catlow and G. Sankar. Determining the structure of active sites, transition states and intermediates in heterogeneously catalysed reactions, *Chem. Commun.*, 2921 (2002).

20. (a) R. D. Oldroyd, G. Sankar, J. M. Thomas and D. Ozkaya. Enhancing the performance of a supported titanium epoxidation catalyst by modifying the active center, *J. Phys. Chem. B*, **102**, 1849 (1998). (b) J. M. Thomas, G. Sankar, M. C. Klunduk, M. P. Attfield, T. Maschmeyer, B. F. G. Johnson and R. G. Bell. The identity in atomic structure and performance of active sites in heterogeneous and homogeneous, titanium-silica epoxidation catalysts, *J. Phys. Chem. B*, **103**, 8809 (1999).

21. J. Jarupatrakorn and T. D. Tilley. Silica-supported, single-site titanium catalysts for olefin epoxidation. A molecular precursor strategy for control of crystal structure, *J. Am. Chem. Soc.*, **124**, 8380 (2002).

22. K. L. Fujdala and T. D. Tilley. Design and synthesis of heterogeneous catalysts: the thermolytic molecular precursor approach, *J. Catal.*, **216**, 265 (2003).

23. I. J. Drake, K. L. Fujdala, A. T. Bell and T. D. Tilley. Dimethyl carbonate production via the oxidative carbonylation of methanol over Cu/SiO_2 catalysts prepared via molecular precursor grafting and chemical vapor deposition approaches, *J. Catal.*, **230**, 14 (2005).

24. A. T. Bell. The impact of nanoscience on heterogeneous catalysis, *Science*, **299**, 1688 (2003).

25. I. J. Shannon, T. Maschmeyer, R. D. Oldroyd, G. Sankar, J. M. Thomas, H. Perrot, J. P. Balikdjian and M. Che. Metallocene-derived, isolated Mo^{IV} active centres on mesoporous silica for the catalytic dehydration of methanol, *J. Chem. Soc. Faraday Trans.*, **94**, 1495 (1998).

26. V. Dal Santo, F. Liguori, C. Perovano and M. Guidotti. Design and use of nanostructured single-site heterogeneous catalysts for the selective transformation of fine chemicals, *Molecules,* **15**, 3829 (2010).

27. D. A. Ruddy and T. D. Tilley. Highly selective olefin epoxidation with aqueous H_2O_2 over surface-modified Ta SBA 15 prepared via the TMP method, *Chem. Commun.*, 3350 (2007).

28. H. C. L. Abbenhuis and R. A. van Santen. From 'nature' to an adventure in single-site epoxidation catalysis, in *Turning Points in Solid-State, Materials and Surface Science,* ed. K. D. M. Harris and P. P. Edwards, RSC Publishing, Cambridge (2007), p.385.

29. H. C. L. Abbenhuis, S. Krijnen and R. A. van Santen. Modelling the active sites of heterogeneous titanium epoxidation catalysts using titanium silsesquioxanes, *Chem Commun.*, 331 (1997).

30. L. Zhang, H. C. L. Abbenhuis, G. Gerritson, N. N. Bhriain, P. C. Magusin, B. Mezari, W. Han, R. A. van Santen, Q. Yang and C. Li. An efficient hybrid, nanostructured, epoxidation catalyst: titanium silsesquioxane-polystyrene copolymer supported on SBA-15, *Eur. J. Chem.*, **13**, 1210 (2007).

31. R. D. Oldroyd, J. M. Thomas, T. Maschmeyer, P. A. MacFaul, D. W. Snelgrove, K. U. Ingold and D. D. W. Wayner. The titanium(IV)-catalyzed epoxidation of alkenes by *tert*-alkyl hydroperoxides, *Angew. Chem. Int. Ed.*, **35**, 2787 (1996).

32. L. Marchese, E. Gianotti, V. Dellarocca, T. Maschmeyer, F. Rey, S. Coluccia and J. M. Thomas. Structure-functionality relationships of grafted Ti-MCM 41 silicas. Spectroscopic and catalytic studies, *Phys. Chem. Chem. Phys.*, **1**, 585 (1999).

33. (a) R. L. Brutchey, D. A. Ruddy, L. K. Andersen and T. D. Tilley. Influences of surface modification of Ti-SBA 15 catalysts on the epoxidation mechanism of cyclohexene with aqueous H_2O_2, *Langmuir,* **21**, 9576 (2005).

    (b) M. Guidotti, C. Pirovano, N. Ravasio, B. Lazaro, J. M. Fraile, J. A. Mayoral, B. Coq and A. Galarneau. The use of H_2O_2 over titanium-grafted mesoporous silica catalysts: a step further towards sustainable epoxidation, *Green Chem.*, **11**, 1421 (2009).

34. N. Ravasio, F. Zaccheria, M. Guidotti and R. Psaro. Mono- and bifunctional heterogeneous catalytic transformation of terpenes and terpenoids, *Top. Catal.*, **27**, 157 (2004).

35. M. Guidotti, N. Ravasio, R. Psaro, G. Ferranis and G. Moretti. Epoxidation on titanium-containing silicates: do structural features really affect the catalytic performance?, *J. Catal.*, **214**, 242 (2003).

36. M. Guidotti, N. Ravasio, R. Psaro, E. Gianotti, L. Marchese and S. Coluccia. Heterogeneous catalytic epoxidation of fatty acid methyl esters on Ti-grafted silicas, *Green Chem.*, **5**, 421 (2003).

37. M. Guidotti, N. Ravasio, R. Psaro, E. Gianotti, S. Coluccia and L. Marchese. Epoxidation of unsaturated FAMEs obtained from vegetable source over Ti$^{(IV)}$-grafted silica catalysts: a comparison between ordered and non-ordered mesoporous materials, *J. Mol. Catal.*, **250**, 218 (2006).

38. E. Gianotti, C. Bisio, L. Marchese, M. Guidotti, N. Ravasio, R. Psaro and S. Coluccia. Ti$^{(IV)}$ catalytic centers grafted on different siliceous materials: spectroscopic and catalytic study, *J. Phys. Chem.*, **111**, 5083 (2007).

39. K. Sato, M. Aoki and R. Noyori. Green oxidation with aqueous hydrogen peroxide, *Chem. Commun.*, 1977 (2003).

40. P. T. Anastas and M. M. Kirchhoff. Origin, current status and future challenges of green chemistry, *Acc. Chem. Res.*, **35**, 686 (2002).

41. F. D. Gunstone. Lipid chemistry: a personal view of some developments in the last 60 years, *Biochim. Biophys. Acta*, **1631**, 207 (2003).

42. J. M. Thomas, R. Raja and D. W. Lewis. Single-site heterogeneous catalysts, *Angew. Chem. Int. Ed.*, **44**, 6456 (2005).

43. J. M. Thomas and R. Raja. The advantages and future potential of single-site heterogeneous catalysts, *Top. Catal.*, **40**, 3 (2006).

44. J. M. Thomas and R. Raja. Designing catalysts for clean technology, green chemistry and sustainable development, *Annu. Rev. Mater. Sci.*, **35**, 315 (2005).

45. O. A. Kholdeeva, I. D. Ivanchikova, M. Guidotti and N. Ravasio. Highly efficient production of 2,3,5-trimethyl-1,4-benzoquinone using aqueous H_2O_2 and grafted Ti(IV)/SiO_2 catalyst, *Green Chem.*, **9**, 731 (2007).

46. O. A. Kholdeeva, V. N. Romannikov, N. N. Trukham and V. N. Parmon. Russian Patent No. 2196764 (2001).

47. O. A. Koldeeva, I. D. Ivanchikova, M. Guidotti, C. Perovano, N. Ravasio, M. V. Barmatova and Y. A. Chesalov. Highly selective oxidation of alkylphenols to benzoquinones with H_2O_2 over silica-supported Ti catalysts: titanium cluster site versus titanium single site, *Adv. Synth. Catal.*, **351**, 1877 (2009).

48. D. Veljanovski, A. Sakthivel, W. A. Herrmann and F. E. Kuhn. Heterogenization of acylperrhenate on mesoporous materials and its applications, *Adv. Synth. Catal.*, **348**, 1752 (2006).

49. R. L. Brutchley, I. J. Drake, A. T. Bell and T. D. Tilley. Liquid-phase oxidation of alkylaromatics by a H-atom transfer mechanism with a new heterogeneous Co SBA-15 catalyst, *Chem. Commun.*, 3736 (2005).

50. D. Bek, N. Zilkova, J. Dedecek, J Sedlacek and H. Balcar. SBA-15 immobilized ruthenium carbenes as catalysts for ring closing metathesis and ring opening metathesis polymerization, *Top. Catal.*, **53**, 200 (2010).

51. C. W. Jones. On the stability and recyclability of supported metal-ligand complex catalysts: myths, misconceptions and critical research needs, *Top. Catal.*, **53**, 942 (2010).

52. J. M. Thomas. Heterogeneous catalysis: enigmas, illusions, challenges, realities, and emergent strategies of design, *J. Chem. Phys.*, **128**, 182502 (2008).

53. N. N. Tusar, S. Jank and R. Gläser. Manganese-containing porous silicates: synthesis, structural properties and catalytic applications, *Chem. Cat. Chem.*, **3**, 254 (2011).

54. J. M. Thomas and R. Raja. Mono-, b- and multifunctional single-sites: exploring the interface between heterogeneous and homogeneous catalysis, *Top. Catal.*, **53**, 848 (2010).

55. M. Iwamoto and Y. Kosugi. Highly selective conversion of ethene to propene and butenes on nickel ion-loaded mesoporous silica catalysts, *J. Phys. Chem. C*, **111**, 13 (2007).

56. M. Tanaka, A. Itadani, Y. Kuroda and M. Iwamoto. See Abstracts of Europacat IX, Salamanca, paper 02-13 (2009), p.129.

57. M. Choi, F. Kleitz, D. Liu, H. Y. Lee, W. S. Ahn and R. Ryoo. Controlled polymerization in mesoporous silica toward the design of organic-inorganic composite nanoporous materials, *J. Am. Chem. Soc.*, **127**, 1924 (2005).

58. P. P. Pescarmona, J. C. van der Waal, I. E. Maxwell and T. Maschmeyer. A new, efficient route to titanium-silsesquioxane epoxidation catalysts developed by using high-speed experimentation techniques, *Angew. Chem. Int. Ed.*, **40**, 740 (2001).
59. F. Carniato, C. Bisio, E. Boccaleri, M. Guidotti, E. Gavrilova and L. Marchese. Titanosilsesquioxane anchored on mesoporous silicas: a novel approach to the preparation of heterogeneous catalysts for selective oxidations, *Chem. Eur. J.*, **14**, 8098 (2008).
60. A. Taguchi and F. Schüth. Ordered mesoporous materials in catalysis, *Micropor. Mesopor. Mat.*, **77**, 1 (2005).
61. F. Hoffmann and M. Fröba. Vitalising porous inorganic silica networks with organic functions: PMO and related hybrid material, *Chem. Soc. Rev.*, **40**, 608 (2011).
62. A. Karkamkar, S. S. Kim and T. J. Pinnavaia. Hydrothermal restructuring of the cell and window sizes of silica foams, *Chem. Mater.*, **15**, 11 (2003).
63. T. Sen, G. J. T. Tiddy, J. L. Casci and M. W. Anderson. Macro-cellular silica foams: synthesis, *Chem. Commun.*, 2183 (2003).
64. P. Waller, Z. Shan, L. Marchese, G. Tartaglione, W. Zhou, J. C. Jansen and T. Maschmeyer. Zeolite nanocrystals inside mesoporous TUD-1: a high-performance catalytic composite, *Chem. Eur. J.*, **10**, 4970 (2004).
65. J. A. Ballantine, J. H. Purnell and J. M. Thomas. Sheet silicates: broad spectrum catalysts for organic synthesis, *J. Mol. Cat.*, **27**, 157 (1984).
66. J. A. Ballantine, M. Davies, H. Purnell, M. Rayanakorn, J. M. Thomas and K. J. Williams. Chemical conversions using sheet silicates: facile ester synthesis by direct addition of acids to alkenes, *J. Chem. Soc. Chem. Commun.*, 8 (1981).
67. J. M. Thomas. Sheet silicate intercalates: new agents for unusual chemical conversions, in *Intercalation Chemistry*, ed. M. S. Whittingham and A. J. Jacobson, Academic Press, New York (1982), p.56.
68. P. A. Diddams, J. M. Thomas, W. Jones, J. A. Ballantine and J. H. Purnell. Synthesis, characterization and catalytic activity of beidellite-montmorillonite silicates and their pillared analogues, *Chem. Commun.*, 1340 (1984).
69. W. Jones. Utilizing clays and other layered solids for the design of new materials, *University of Wales Review: Science and Technology*, **2**, 45 (1985).

70. J. M. Thomas. Solid acid catalysts, *Sci. Am.*, **266**, 85 (1992).

71. J. M. Thomas. New ways of characterizing layered silicates and their intercalates, *Phil. Trans. R. Soc. A*, **311**, 271 (1984).

72. J. A. Ballantine, J. H. Purnell and J. M. Thomas. Method of producing esters utilizing the proton-rich nature of the interlamellar spaces of montmorillonite, U.S. Patent No. 4899319 (1985) and E.P. Patent No. 31252 (1984).

73. (a) W. J. Roth and J. Čejka. Two-dimensional zeolites: dream or reality, *Catal. Sci. Tech.*, **1**, 43 (2011).

    (b) D. Liu, A. Bhan, M. Tsapatsis and S. Al-Hashimi. Catalytic behaviour of Brønsted acid sites in MWW and MFI zeolites with dual meso- and microporosity, *A. C. S. Catal.*, **1**, 7 (2011).

74. C. R. A. Catlow, S. A. French, A. A. Sokol and J. M. Thomas. Computational approaches to the determination of active site structures and reaction mechanisms in heterogeneous catalysis, *Phil. Trans. R. Soc. A*, **363**, 913 (2005).

75. A. Corma and H. Garcia. Lewis acids as catalysts in oxidation reactions: from homogeneous to heterogeneous systems, *Chem. Rev.*, **102**, 3837 (2002).

76. (a) J. M. Thomas, J. C. Hernandez-Garrido, R. Raja and R. G. Bell. Nanoporous oxidic solids: the confluence of heterogeneous and homogeneous catalysis, *Phys. Chem. Chem. Phys.*, **11**, 2799 (2008).

    (b) I. Luz, F.X.L. i Xamena and A. Corma. Bridging homogeneous and heterogeneous catalysis with MOFs: "Click" reactions with Cu-MOF catalysts, *J. Catal.*, **276**, 134 (2010).

77. H. Gao and R. J. Angelici. Combination catalysts consisting of a homogeneous catalyst tethered to a silica-supported Pd heterogeneous catalyst: arene hydrogenation, *J. Am. Chem. Soc.*, **119**, 6937 (1997).

78. J. M. Thomas. The advantages of exploring the interface between heterogeneous and homogeneous catalysis, *Chem. Cat. Chem.*, **2**, 127 (2010).

CHAPTER 7

EXPLOITING NANOSPACE FOR ASYMMETRIC CONVERSIONS

7.1 Background

A central feature of modern chemistry is the desire to increase the variety of ways in which enantiomerically pure compounds may be prepared. The demand for chirally pure materials has never been greater: pharmaceuticals, agrochemicals and many fragrances and flavours need to be prepared in enantiopure form. In consequence, the so-called enantiomeric excess (*ee*) associated with a given conversion is the figure of merit that is universally used. The *ee* of a reaction is defined by

$$ee = [P_R] - [P_S]/[P_R] + [P_S]$$

where $[P_{R/S}]$ stands for the concentration of the product in its R or S enantiomer. Values of *ee* are always quoted as a positive percentage.

Traditionally in organic chemical synthesis it has been the practice to carry out asymmetric conversions in homogeneous solutions usually by utilizing soluble chiral catalysts, many of them being organometallic or metal complex species. Such procedures are not ideal since, in separating the desired product from unreacted starting material, there is an inevitable loss of catalyst. And since both the central metal (e.g. Rh or Pt) is itself not cheap, and since the chiral ligand which is attached to it is often even more expensive than the metal, this is usually a costly operation. The incentive to design chiral solid catalysts is therefore considerable. In this chapter we shall focus on three main kinds of chiral nanoporous solids that are suitable for asymmetric catalysis. An attractive feature of enantioselective solid catalysts lies in the effect of chiral multiplication: minute amounts of the catalyst can, ideally, produce large amounts of chiral compounds.

7.2 Whither Chiral Zeolites?

Ever since the late Richard Barrer,[1] a pioneer of zeolite science, demonstrated 50 years ago the relative case of synthesizing zeolites, it has been a gleam in the eye of many academics and technologists to synthesize chiral microporous and mesoporous crystalline solids. The applications of such solids are tantalizingly exciting: facile separation of enantiomeric molecules; the synthesis of enantiopure isomers; and the prospect of rapid separations by liquid chromatography are but a few of the possibilities. But such chiral, crystalline, open-structure solids are hard to come by. Of the 180 or so known zeolite structures, less than a handful of these exhibit chiral frameworks, the best known being polymorph A of zeolite Beta.[2] Imaginative efforts in Stockholm by Zou *et al.*,[3] and elsewhere, have produced two new chiral zeolites: the so-called SU-32 and SU-15 silicogermanates.

Recent discussions (by Harris and Thomas[4] and by Morris and Bu[5]) have highlighted the fact that succeeding in crystallizing a zeolite in a chiral space group — of which there are 65 out of the total number of 230 — is not enough. To be of any value in asymmetric catalysis, the sample of a prepared chiral zeolite must be homochiral: it must all be of the same handedness. Zeolitic catalysis is almost invariably carried out using polycrystalline samples, rather than a single crystal. Figure 7.1 illustrates the key importance of having a homochiral sample.

One must also avoid crystallographic twinning, which, if present, can convert a chiral crystal into one possessing feeble chirality. (A classic case, seen in the work of Ramdas *et al.*[6] concerns the "chiral" solid hexahelicene.[7])

Many inorganic materials have chiral crystal structures, but they often form only a racemic conglomerate that consists of both the left-handed and right-handed variants in the same solid. The classic

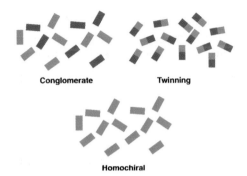

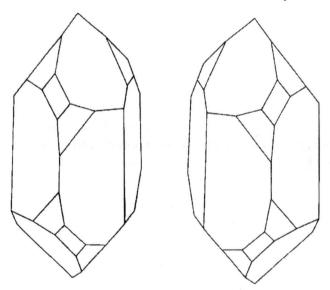

Figure 7.1 Materials that crystallize with a chiral structure may give rise to various possibilities in a bulk sample. The top two illustrations are each non-chiral because they have equal amounts of the right-handed (red) and left-handed (blue) crystals, or parts of crystals. In aiming to prepare a strictly chiral crystal (of zeolite or AlPO) one should endeavour to form a homochiral (bottom) bulk sample.[4,14]

Figure 7.2 Schematic illustration of the enantiomorphous crystal morphologies of D-quartz and L-quartz.[4]

example is quartz (see Figure 7.2) which so intrigued Lord Kelvin,[8] who coined the word "chiral". (Another is sodium chlorate which stimulated the curiosity of Lord Rayleigh and the remarkable Italian stereochemist Perucca.[9])

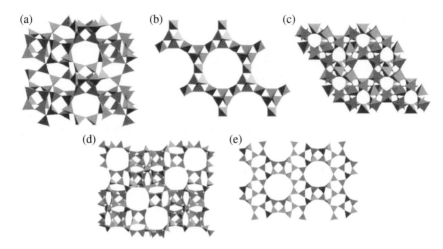

Figure 7.3 A selection of chiral zeolites. There are now many zeotype structures that have been identified as being chiral. This figure shows a polyhedral representation of several of the potentially most interesting chiral zeotypes: (a) SU-32, (b) OSB-1, (c) CZP, (d) ITQ-37. Although these materials all crystallize in chiral space groups they have not yet been prepared as homochiral bulk solids (except for the zincophosphate form of zeolite CZP). (e) The predicted structure of polymorph A of zeolite Beta, which has yet to be isolated in pure form — it always tends to form intergrowths with other polymorphs. (After Morris and Bu.[5])

Ever since it was recognized[2] that zeolite Beta, which is of great importance in shape-selective catalysis as we saw in Chapter 4, has a potential chiral polymorph, there has been much effort devoted to the preparation of chiral zeolites and AlPOs.[10–14] Five examples of zeolite frameworks identified as chiral are shown in Figure 7.3.

Many attempts have been made to prepare further examples of such chiral microporous solids. The only successful strategy seems to be that of Parnham and Morris[15] using the so-called ionothermal approach which entails the use of ionic liquids.

No one has yet unequivocally shown that a chiral zeolite — scarce as they are at present — can lead to useful asymmetric catalysis. It is rather disappointing that none of the very few chiral zeolites (e.g. ITQ-37) has led to significant advances in this direction. Harris and Thomas[4] have emphasized the importance of recognizing that the

ability of a chiral host structure to exert control over the enantioselectivity of asymmetric reactions occurring within it relies on the two
enantiomeric pathways within a given enantiomorphous form of the
host structure being associated with a sufficiently large difference in
energy. This is a factor also discussed by Davis[16] and by Willock *et al.*[17]

A fascinating paper by Yu and Xu[12] described a heteroatom-
containing AlPO helical structure, designated MAPO-CJ40 in which
M = Co, Zn, that crystallizes in the chiral space group $P2_12_12_1$. The
framework structure is illustrated in Figure 7.4. Interestingly, they

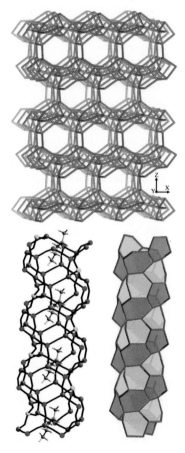

Figure 7.4 (a) Framework structure of Co AlPO-CJ40 viewed along the *b*-axis; (b) helical channel with a ten-ring window; and (c) double helical chains around the ten-ring
channel made of MO_4 (M = Co and/or Al) and PO_4 tetrahedra. (After Yu and Xu.[12])

established that the Co^{II} ions (replacing some Al^{III} ones) distributed along the wall of the channels show a helical arrangement. The circular dichroism measurements exhibit a strong Cotton effect, suggesting that the resultant crystals are not racemic, even though the synthesis did not involve any chiral starting materials. The helical channels in this novel MAPO structure run along the <010> directions; they have ten-ring openings with a pore size of 4.4 × 2.2 Å. It is conceivable that, with the appropriately chosen prochiral substrate, this single-site chiral heterogeneous catalyst could be a prime candidate for an experimental proof of zeolite asymmetric catalysis. Difficulties arise, however, because once the template is removed from MAPO-CJ40, it collapses.

Some progress in rendering zeolites enantioselective in their catalysed conversions has been made by Hutchings[18] who introduced chiral organic "guest" species into the micropores to steer the reaction in a chirally preferred manner. These workers produced an SSHC capable of effecting asymmetric catalysis by modifying zeolite Y with dithiane oxide so as to engender novel enantioselectivity. It transpires that when zeolite Y is modified by an enantiomerically enriched dithiane 1-oxide, the resulting solid catalyst is selective in the dehydration of one of the enantiomers of butan-2-ol, even though both enantiomers are present in equal concentration at the inlet of the reactor.

7.3 Chiral Metal-organic Frameworks (MOFs)

In earlier chapters we alluded to the fact that, because of their zeolite-mimicking features relating to size and shape selectivity, MOFs are potentially useful candidates as materials for the design of SSHCs.

By taking advantage of the mild conditions that are typically used for the synthesis of MOFs, Wu and Lin[19] and others[5,20–22] have successfully synthesized catalytically active homochiral MOFs by incorporating chiral constituent building blocks that contained orthogonal functionalities. In 2007, Wu and Lin produced two closely related homochiral nanoporous MOFs built from the same

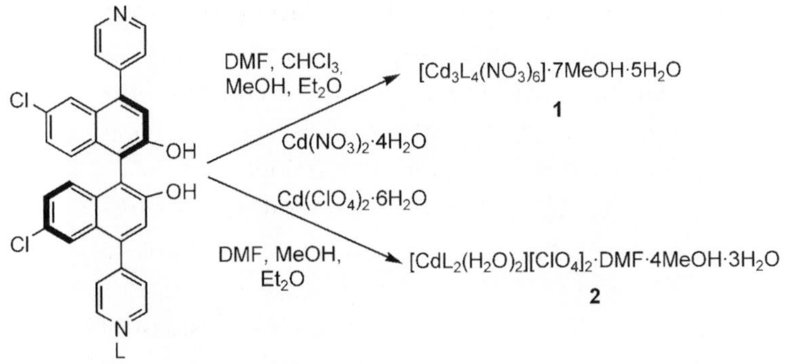

Figure 7.5 Outline of the synthesis (by Wu and Lin[19]) of two similar chiral MOFs, **1** and **2** (see text).

chiral bridging ligand and metal connecting point. However, for reasons that are structurally intelligible, only one of these exhibited the ability to effect asymmetric catalysis. To understand why, we first consider (Figure 7.5) the procedure for producing the two MOFs **1** and **2**.

Solid **1** is shown in greater detail in Figure 7.6. The heterogeneous catalyst was derived from **1** by activating Lewis acidic metal centres (namely Ti(OiPr)$_4$) with the chiral dihydroxy groups that are present as the orthogonal secondary functionalities in solid **1**. It is well known (in homogeneous systems) that Ti^{IV}-BINOLate complexes are active catalysts for a range of organic transformations that are typically catalysed by Lewis acids. Treatment of **1** with excess Ti(OiPr)$_4$ in toluene led to an active catalyst for the addition of diethylzinc to aromatic aldehydes to yield chiral secondary alcohols. Specifically, this Ti-**1** SSH catalyst led to the addition of diethylzinc to 1-naphthaldehyde to produce (R)-1-(2-naphthyl)-propanol with complete conversion and 90% *ee* — see the first entry in Table 7.1.

The reason why the not entirely dissimilar material **2** is not catalytically active is because, in its crystalline structure, the pyridyl and naphthyl rings form mutually perpendicular interpenetrating two-dimensional (2D) rhombic grids from π...π interactions with a nearest C...C separation of 3.27 Å. As a consequence, all the chiral dihydroxy groups of the L ligands (Figure 7.5) are held so closely to

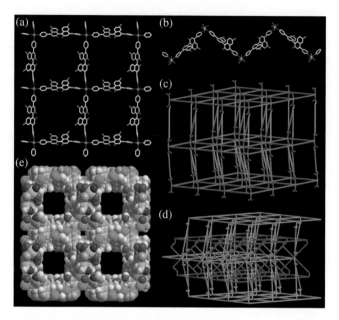

Figure 7.6 Crystal structure of compound **1** (see Figure 7.5). (a) The 2D square grid in **1**. (b) The 1D zigzag polymeric chain in **1**. (c) Schematic representation of the 3D framework of **1**. (d) Schematic representation of the twofold interpretation of **1**. (e) Space-filling model of **1** as viewed down the *c*-axis showing the chiral 1D channels of 13.5×13.5 Å^2 in dimensions. Colour scheme: (a), (b) and (e): cyan, Cd; green, Cl; red, O; blue, N; grey, C; and light grey, H. (After Wu and Lin.[19])

Table 7.1 Addition of diethylzinc to aromatic aldehydes.[a]

Ar	Ligand	Conversion [%]	*ee* [%]
1-Naphthyl	1	> 99	90.0
4-CH$_3$Ph	1	> 99	84.2
4-CH$_3$Ph	1	> 99	84.9
Ph	1	> 99	81.9
4-ClPh	1	> 99	60.2

[a] All reactions were conducted at room temperature for 15 h with **1** or L and excess amounts of Ti(O*i*Pr)$_4$ in toluene.

the $[Cd(py)_2(H_2O)_2]$ hinges that there is too much steric congestion around the dihydroxy groups so that the substitution of two isopropoxide groups by the BINOLate functionality is precluded.

The message in this early work of Lin's is that MOFs offer a good platform for designing asymmetric SSHCs. All active sites have the same environment and are spatially far apart: they are all located in a chiral environment. In a more recent paper,[23] Lin and his team have engineered systematically chiral MOFs (CMOFs) with tunable functionalities and large open channels. Figure 7.7 illustrates the nature

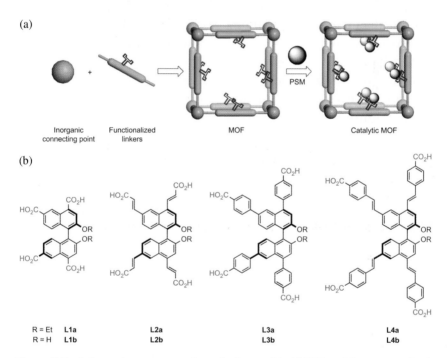

Figure 7.7 Schematic representation of a homochiral MOF and its post-synthetic modification (PSM) to give a catalytically active MOF. (a) The assembly of CMOFs and their PSM to give heterogenized asymmetric catalysts. (b) The chemical structures of the ligands used for this study. The notations **L1a–L4a** and **L1b–L4b** are used to describe both the protonated (as in free ligands) and deprotonated forms (as in CMOFs) of (R)-tetracarboxylic acid. R = Et for CMOF-**1a** to -**4a** and R = H for CMOF-**1b** to -**4b**. The primary carboxylic acid groups highlighted in red are used to form CMOFs, whereas the secondary dihydroxy groups highlighted in blue react with Ti(OiPr)$_4$ on the PSM to form asymmetric catalysts.

and assembly of a family of eight mesoporous homochiral CMOFs with the framework formula [LCu$_2$(solvent)$_2$], where L is a chiral tetracarboxylic ligand derived from 1,1′-bi-2-naphthol. All these have the same structure but different diameters of pores. Chiral Lewis acid catalysts were again generated by post-synthesis functionalization with Ti(O*i*Pr)$_4$. The resulting solids proved to be highly active asymmetric catalysts for converting aromatic aldehydes into chiral secondary alcohols, using diethylzinc and alkynylzinc additions, as shown, along with relevant X-ray powder patterns in Figure 7.8. Conversions of greater than 95%, and *ee* values well over 80% were achieved. The channel diameters reported in this work range from 8 Å to 21 Å, with void spaces falling in the range of 73% to 92%. Such impressive laboratory-based achievements have not, to our knowledge, been exploited industrially. More developments are expected in SSHCs involving MOFs, both for enantio- and non-enantioselective conversions in future years (see especially Refs 23(b) and (c)).

7.4 Harnessing the Asymmetric Catalytic Potential of Mesoporous Silicas Using SSHCs

7.4.1 *Background*

In the general trend to combine the advantages of homogeneous and heterogeneous catalysis for effecting enantioselective conversions, many approaches involving the heterogenization of chiral metal complexes have been tried.[24] Some well-known examples involve the following:

- chiral Mn^{III} salen complexes encapsulated in zeolite Y for epoxidation of aromatic alkenes[25] in the presence of NaOCl;
- Mo^{VI} hydroxyproline complexes grafted onto zeolites for the epoxidation of allylic alcohols[26];
- Os^{VIII} (9-*O*-quininyl)phthalazine complexes for the dihydroxylation of olefins[27]; and
- Rh^{I}-BINAP encapsulated in the layered aluminosilicate hectorite for the hydrogenation of α,β-unsaturated esters.[28]

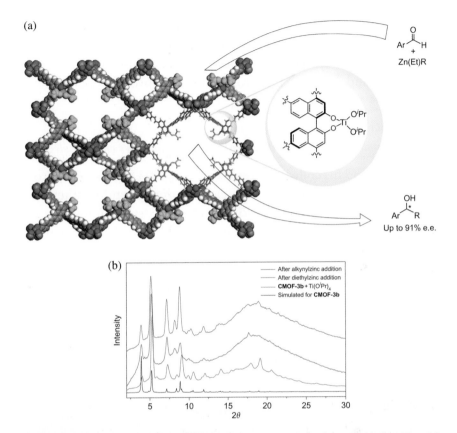

Figure 7.8 CMOF-derived asymmetric catalysts and their framework stability. (a) Schematic representation of asymmetric alkyl- and alkynylzinc additions catalysed by the CMOF/Ti-BINOLate catalyst within large open channels. Inductively coupled plasma mass spectroscopy analyses indicate binding of Ti^{IV} to the CMOF framework. (b) PXRD patterns of CMOF-**3b**, Ti(OiPr)$_4$-treated CMOF-**3b** and recovered CMOF-**3b**/Ti(OiPr)$_4$ catalysts after diethylzinc and alkynylzinc addition reactions. The identical PXRD patterns between these samples indicate that the framework structure of CMOF-**3b** is maintained after Ti^{IV} loading and after catalytic diethylzinc and alkynylzinc addition reactions.

None of these, however, turned out to be wholly successful so far as stability, reproducibility and TOFs were concerned. Many of them ceased to effect enantioselectivity after rather meagre use. One problem appeared to be the dimerization, or other accretion, of the active species at the solid surface to which the complexes were anchored. Nevertheless, they were generally superior to chirally modified metals, typified by adsorbed cinchonidine and its derivatives on Ni and other surfaces for asymmetric hydrogenation.[24]

7.4.2 *Exploiting nanospace for asymmetric catalysis: confinement of immobilized single-site chiral catalysts enhances enantioselectivity*

The author and his colleagues realized[29,30] in 1995, when mesoporous silicas had by that time proved so easy to prepare and characterize, that unique opportunities would henceforth exist to capitalize, catalytically, on engineered active centres that operated in a spatially constrained manner, as schematized in Figure 7.9. The idea here is that a high-area, achiral silica surface, rich in pendant Si-OH groups (as in Figure 7.9(a)), be used as a support for a chiral organometallic entity in such a manner as to restrict the approach of a prochiral molecule such as a ketoester, that is to be hydrogenated

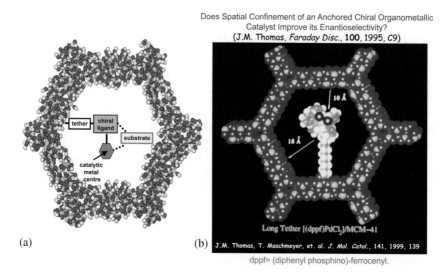

(a)   (b)

Does Spatial Confinement of an Anchored Chiral Organometallic Catalyst Improve its Enantioselectivity?
(J.M. Thomas, *Faraday Disc.*, **100**, 1995, C9)

10 Å

18 Å

Long Tether [(dppf)PdCl₂]/MCM–41

J.M. Thomas, T. Maschmeyer, et. al. *J. Mol. Catal.*, 141, 1999, 139

dppf= (diphenyl phosphino)-ferrocenyl.

Figure 7.9   Enantioselectivity and asymmetric syntheses from the exploitation of molecular confinement in nanoporous solids. (After Thomas[29] and Thomas and Raja.[43])

(Figure 7.9(b)) via the Pd active centre. By matching the length of the tether and the diameter of the nanoporous silica, the extra confinement imposed upon the prochiral species in reaching the active site boosts the enantioselectivity well beyond that for the same chiral complex anchored on a flat or convex silica surface. The deliberate restriction of spatial freedom for the reactant in the vicinity of the active site results in additional interactions of the reactant with the pore wall. This interaction would be approximately equal to the energy difference between the two transition states that lead to the chiral products.

In practice — if this notion is valid — a prochiral olefin, for example, approaching a single-site chiral catalyst such as a Pd complex with diphenylphosphino-ferrocenyl (dppf) would, upon hydrogenation, exhibit enhanced *ee* compared with the value obtained when the same chiral catalyst functioned in homogeneous solution. Not only would chiral hydrogenations be boosted, so would a multiplicity of other conversions in the wide realm of organic chemistry: for example, aminations, carbamate syntheses and epoxidations using the appropriate immobilized chiral catalyst.

The first practical test for this strategy involved C-N bond formation in the allylic amination (the Trost–Tsuji reaction[31]) of cinnamyl acetate (see Figure 7.10). Three different catalysts were used[32] (two heterogeneous and one homogeneous), but with the same chiral active site in all three (see Figure 7.11). Note, in particular, that the dppf active site is exactly the same when anchored to both the mesoporous and the non-porous silicas and also when attached to

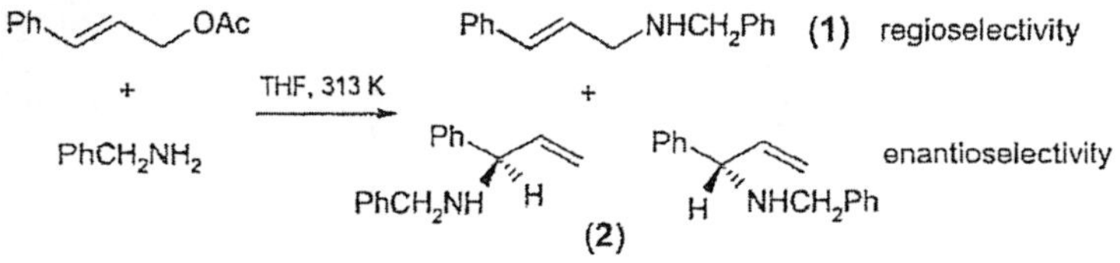

Figure 7.10 The allylic amination of cinnamyl acetate with benzylamine,[31,32] which was used to test the validity of the concept of spatial confinement and its enhancement of enantioselectivity (see text).

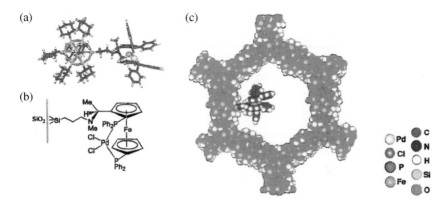

Figure 7.11 Depiction of the catalytically active chiral dppf-diamine palladium dichloride attached to a soluble silsesquioxane moiety (a); tethered to a non-porous silica (Cabosil) (b); and spatially confined within a mesoporous silica (MCM-41) (c).[32,43] (dppf = (S)-1-[(R)-1,2'-bis(diphenylphosphino)ferrocenyl] which is covalently tethered to the silsesquioxane, the mesoporous silica and Cabosil with ethyl-N,N-dimethylethylenediamine.)

the soluble silsesquioxane (Si$_8$O$_{12}$) which has cyclohexyl groups attached to seven of its eight vertices. The chiral ligand attached to the dppf PdCl$_2$, which in turn was covalently tethered to the silicas, was ethyl-N,N'-dimethylethylenediamine. The confined catalyst used in this proof-of-principle experiment was tethered inside *ca* 30 Å diameter mesopores in an MCM-41 silica. The unconfined catalyst consisted of Cabosil, a non-porous commercial silica, to which was tethered the same chiral (S)-1-(R)-1,2'-bis(dppf)ethyl-N,N'-dimethylenediamine Pd Cl$_2$ active centre. (To ensure that no active sites are bound to the exterior surface of the mesoporous silica, pre-treatment with diphenyl dichlorosilane was carried out so as to convert all the exposed Si-OH groups initially resident there into chemisorbed phenylated silane.)

The results of this test, chosen to probe simultaneously the regio- and enantioselectivity of the process, show that the unconfined catalyst yields a minute (2%) amount of the desirable branched product, whereas the confined one has a 50% yield of essentially equal amounts of chiral (branched) and straight-chain product. The homogeneous conversion, using the dppf [*c*C$_5$H$_9$Si]$_7$O$_{12}$ moiety, which is soluble in hexane, yields essentially zero *ee*. More significantly, in the confined catalyst, the product exhibits an *ee* value of 95%, far in excess of the *ee* value seen in the minute amount of branched product.

The general feasibility of effecting asymmetric conversions in this way, which has been described in numerous publications from the author's group,[34–37] has been multiply attested by us, by other workers (notably Li *et al.*[38] in China; van Santen[39] in the Netherlands; and Song *et al.*[40,41] in Korea) and by German industrial chemists. In extensive investigations by Thomas *et al.*[32–37,42,43] of the exploitation of nanospace for asymmetric catalysts, the chiral single sites were initially covalently anchored to the siliceous nanopores via a ligand attached to Pd^{II} or Rh^I centres. Later Thomas, Raja and Johnson *et al.* used a more convenient and cheaper electrostatic method, similar to the one used by de Rege *et al.*[44] to immobilize complex organometallic catalysts. This relies in part on strong hydrogen bonding (see Figures 7.12 and 7.13).

Numerous examples of the success of this approach to the construction of effective SSHCs that effect asymmetric conversions have been summarized by Thomas and Raja[43]; and two typical examples are shown in Figure 7.14. The chiral synthesis shown in the lower half of this figure, the hydrogenation of methyl benzoylformate to methyl mandelate, is a commercially important process, since the product is used as a precursor in the synthesis of pemoline, a stimulant of the central nervous system.

Rather than employing the more costly phosphorus-based ligands extensively used to confer chirality in organometallic (precursor) catalysts, the author and his colleagues used the less costly but effective chiral diamine ligands such as variants of substituted diphenylethylenediamine (see Figure 7.13). My group also used a sequence of well-defined, commercially available, nanoporous silicas, each of which has a sharply defined pore diameter.[45] The results pertaining to the influence of the spatial restriction on catalytic performance in the production of methyl mandelate are shown in Figure 7.15. The most striking feature here is the expected sharp diminution in degree of enantioselectivity as spatial congestion decreases with increasing pore diameter.

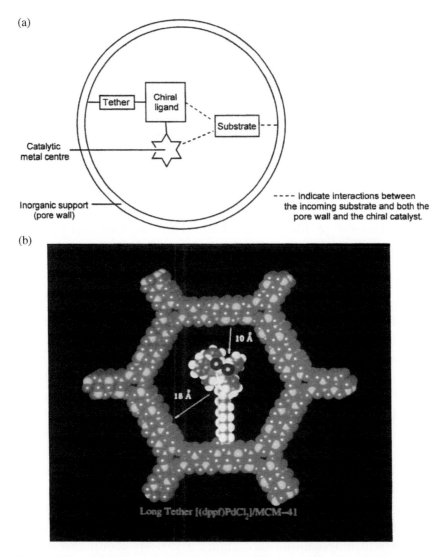

Figure 7.12 (a) Schematic representation of the principle of exploiting spatial confinement to enhance the enantioselectivity of immobilized chiral organometallic catalysts. (b) Attachment of a chiral organometallic Rh^{I}-based cation to a mesoporous silica by means of strong hydrogen bonding involving the silanol groups of the silica and a triflate counterion.[29,43]

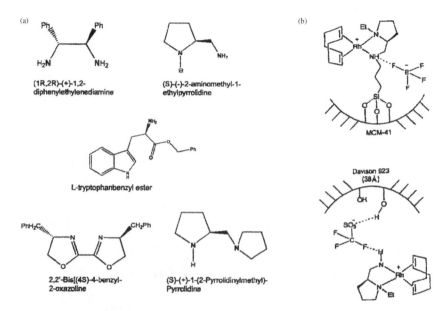

Figure 7.13 Some asymmetric diamine ligands (a) that can be anchored to the concave surfaces of mesoporous silicas via covalent (using BF_4^- anion) and non-covalent methods (using $CF_3SO_3^-$ counterion (b)).[43]

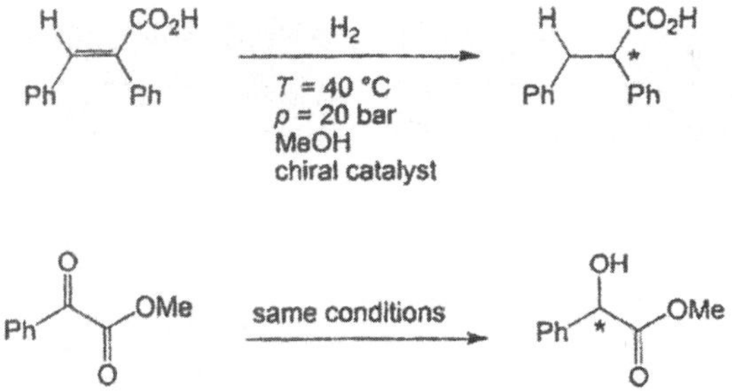

Figure 7.14 Examples of successful enantioselective hydrogenations achieved with single-site organometallic catalysts anchored on achiral surfaces.

The calculated turnover frequencies (ToF in h^{-1}) for these catalysts are 3,154 (38 Å), 3,149 (60 Å) and 3,156 (250 Å), respectively, confirming that the turnover at the active sites is not diffusion-limited. The *ee* values were compared under almost identical levels of activity.

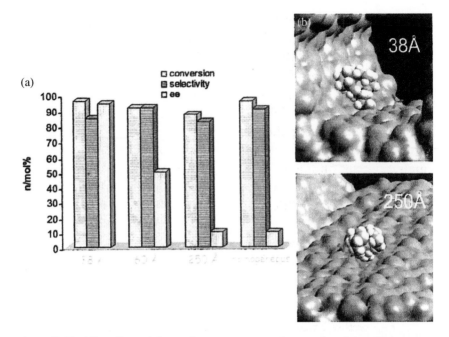

Figure 7.15   The effect of the surface curvature and concavity of the silica support in facilitating the asymmetric hydrogenation of methyl benzoylformate to methyl mandelate using [Rh(COD)-(S)-(+)-1-(2-pyrrolidinylmethyl)-pyrrolidine] $CF_3SO_3^-$ catalysts anchored onto mesoporous silicas of varying pore dimensions.

In view of the great importance and future potential of pursuing asymmetric syntheses and related processes heterogeneously — the practical value of the recyclability alone warrants its further study — we return in Sections 7.4.3 and 7.4.4 to other aspects of the subject.

Similar effects exerted by spatial confinement have been reported by Hutchings *et al.*[46] in the asymmetric epoxidation of (Z)-stilbene with iodosylbenzene as oxygen donor. A detailed study by Zhang and Li[47] investigated the asymmetric epoxidation of 6-cyano-2,2-dimethylchromene also using chiral Mn^{III} (salen) immobilized within mesoporous silica. In particular, they varied the length of the axial linkage separating the Mn^{III} (salen) active site from the wall of the nanopores, the diameters of which were also systematically varied. In harmony with the results of Raja *et al.*,[35,43] dealing with asymmetric hydrogenation (see Figures 7.14 and 7.15), these workers found that the *ee* values increased as the pore diameter decreased and as the length of the linkage increased. For the confined catalyst, *ee* values of 90% were obtained, whereas with the epoxidations in homogeneous solution the *ee* values were *ca* 80%.

An earlier study by Balkus *et al.*[48] investigated the epoxidation of α-methylstyrene, using the same Mn-salen complex as in Ref. 47. These workers found a distinct enhancement of *ee* for the confined catalyst over the homogeneous one: 91% compared with 51%. In a still earlier study, Corma *et al.*,[49] using Mo^{VI} complexes of chiral ligands derived from (2*S*,4*R*)-4-hydroxyproline heterogenized onto ultrastabilized Y zeolite, pre-treated so as to produce mesopores in the matrix, found twice as high an *ee* for the asymmetric oxidation of geraniol than with a homogeneous catalyst.

7.4.3 Asymmetric hydrogenation of E-α-phenylcinnamic acid and methyl benzoylformate: the advantages of using inexpensive diamine asymmetric ligands

Enantiomerically enriched α- and β-hydroxy carboxylic esters are valuable reagents and important intermediates in the preparation of pharmaceuticals and agrochemicals.[50,51] Esters of mandelic acid, for example, are used as stated earlier as precursors in the synthesis of pemoline, a central nervous system stimulant, and in the production of artificial flavours and perfumes. Methyl mandelates are currently prepared by direct esterification of mandelic acid with alcohols in the presence of sulfuric acid.[52] This method of direct esterification with strong acids not only has the disadvantages associated with the corrosive nature of the acid, but is usually accompanied by side reactions such as carbonization, oxidation and etherification, which decreases overall selectivity and enantiopurity.[53] In addition, expensive downstream processing and post-treatment pollution problems make the process commercially and environmentally unattractive.

Even though, as discussed in Chapter 2, enzymes are currently more extensively used industrially than asymmetric transition metal (TM) complexes for enantioselective catalytic conversions involving pharmaceuticals and agrochemicals, TM complexes are of growing importance in this context. Biological processes using micro-organisms for asymmetric reduction have the inherent problem of stability of the enzyme and high costs associated with external co-factors.[51] Organometallic chiral complexes have, however, over-whelmingly, been used[54] homogeneously, and their heterogeneous counterparts have hitherto performed disappointingly so far as their enantioselectivity is concerned, largely because they contained a range of different kinds of active centres, each with its own catalytic selectivity, a situation that contrasts markedly with the one prevailing here with SSHCs that are chiral. Moreover, one of the major disadvantages of using customary chiral phosphines is their high cost — the cost of the ligand, in most applications, outweighs the cost of the final product — and sensitivity to oxidation, which limits their industrial applicability.

Comparatively, few reports have hitherto been published in which Rh[I] or Pd[II] asymmetric complexes without phosphine ligands have been used to activate hydrogen, but a growing number employing nitrogen-containing ligands have appeared of late for the purpose of asymmetric catalysis.[55] The validity of the strategic principle, outlined in the preceding sections of this chapter, which was confirmed and supported by sectors of the German chemical industry when the principles were tested in a commercial context,[56,57] led the author and his colleagues to evaluate the feasibility of tethering rather inexpensive diamine asymmetric ligands (see Figure 7.13) onto the inner walls of mesoporous silicas. The argument behind such an approach was to substantiate the concept that the diamine ligands tethered to concave silica surfaces would boost the enantiose-lectivity to a far greater degree, in contrast to having them attached to convex surfaces. As outlined earlier (Section 7.4.2) the idea is that the confining dimensions of the interior of a pore should be able to restrict the possible orientations that a bulky reactant can assume as it approaches a chiral catalytic centre that is attached to a concave pore wall (see Figure 7.16). If a reactant is, as it were, nudged by its surroundings into the right orientation for a stereospecific reaction, then that reaction should proceed enantioselectively.[39]

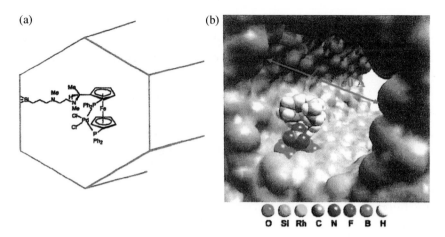

Figure 7.16 Enantioselectivity can be induced by tethering chiral dppf-diamine-based catalysts to the inner walls of mesoporous silica (a) or by anchoring diamine asymmetric ligands (like those shown in Figure 7.13) onto concave surfaces (b). In the case of the former, it is the constraints imposed by the surface surrounding the metal centre, and in the case of the latter, it is the restricted access generated by the concavity of the pore that is the principal determinant for the ensuing enantioselectivity.[43]

The ligands shown in Figure 7.16 can be anchored covalently[36] to the inner walls of the mesoporous silica, but also using a BF_4^- anion that is hydrogen-bonded to the nitrogen of the amino group in the ligand to secure the cationic organometallic catalyst in place. The catalyst itself is pseudo-square-planar where the metal (Rh^I or Pd^{II}) is bonded to 1,5-cyclooctadiene (COD). The results for the hydrogenation of E-α-phenylcinnamic acid and methyl benzoylformate are summarized in Table 7.2. Further, by grafting the Rh^I chiral complex on both a concave silica (using mesoporous MCM-41 of 30 Å diameter) and a convex silica (a non-porous Cabosil), it was established beyond doubt that it is the spatial restrictions imposed by the concave surface at which the active site is located that enhances the enantioselectivity of the catalyst (see Figure 7.17).

All this corroborates the results shown earlier in Figure 7.15.

7.4.4 One step is better than two

Industrial chemists often draw attention to the fact that a major component of the cost of any chemical manufacturing process stems

Single-site Heterogeneous Catalysts

Table 7.2 Asymmetric hydrogenation of E-α-phenyl cinnamic acid and methyl benzoylformate using covalently tethered Rh(I) and Pd(II) chiral catalysts.

Amine and diene	Catalyst	Substrate	Metal	Time (hr)	Conv.	Sel.	ee
	Homogeneous	α-phenyl cinnamic acid	Rh(I)	24	74	77	93
	Heterogeneous			24	80	74	96
	Homogeneous	Methyl benzoyl formate	Rh(I)	24	95	58	0
COD	Heterogeneous			1	98	94	91
	Homogeneous	α-phenyl cinnamic acid	Rh(T)	24	88	76	64
	Heterogeneous			24	99	66	91
	Homogeneous	Methyl, benzoyl formate	Rh(T)	0.5	85	70	18
NBD	Heterogeneous			2	100	98	99

(Continued)

Table 7.2 *(Continued)*

Amine and diene	Catalyst	Substrate	Metal	Time (hr)	Cmv.	Sel.	ee
	Homogeneous	α-phenyl cinnamic acid	Pd(II)	24	100	87	76
	Heterogeneous			24	93	87	93
	Homogeneous	Methyl'benzoyl formate	Pd(II)	2	100	94	0
	Heterogeneous			2	97	97	87
	Homogeneous	α-phenyl cinnamic acid	Rh(T)	24	57	84	81
	Heterogeneous			24	98	80	93
	Homogeneous	α-phenyl cinnamic acid	Pd(II)	24	95	82	79
	Heterogeneous			24	75	100	88

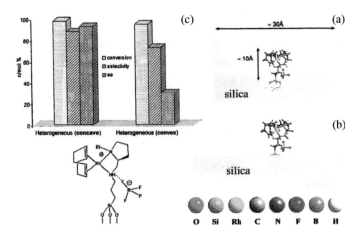

Figure 7.17 Graphical model (to scale) showing the constraints imposed on a Rh^I-(COD)-(S)-(-)-2-aminomethyl-1-ethylpyrrolidine catalyst when it is anchored on a concave (a) and convex surface (b). The computational model shown here has been derived from the crystal structure of the Rh^I complex, which clearly indicates that the BF$_4^-$ anion is hydrogen-bonded to the nitrogen of the amino group. The chiral diamine organometallic catalyst constrained at the concave silica surface (a) surpasses the performance (both in terms of selectivity and *ee*) of the same catalyst anchored to a convex surface (shown in (b)) in the asymmetric hydrogenation of *E*-α-phenylcinnamic acid.[43]

from the necessity to effect various separations. For that reason alone, a process that can be effected in a single step is preferred to those that require two or more separate steps. In this section we focus on a one-step, enantioselective reduction of ethyl nicotinate to ethyl nipecotinate (see Figure 7.18). Each of these materials is of biological significance.[58]

Blaser *et al.*[59] in targeting the enantioselective synthesis of ethyl nipecotinate found it necessary to employ two distinct steps: in the first, the starting material was converted to the 1,4,5,6-tetrahydronicotinate by using a Pd on carbon hydrogenation catalyst. To effect the second and much more demanding step, these workers investigated both unmodified and 10,11-dihydrocinchonidine-modified noble metal catalysts. The best values of activity reported by these workers using the chiral-modified Pd catalyst for the second step was 19% *ee* at 12% conversion.

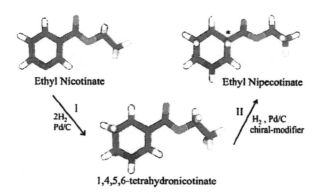

Ethyl Nicotinate   **Ethyl Nipecotinate**

1,4,5,6-tetrahydronicotinate

Figure 7.18 The two steps shown here, each requiring separate heterogeneous catalysts, the second using a cinchonidine modifier, can be carried out in one step using a spatially constrained dppf-based chiral catalyst (see text).[58]

Using the principles of confined chiral catalysis outlined in previous sections of this chapter, Raynor *et al.*[60] were able to achieve a one-step conversion in excess of 50% with an *ee* of 17%. The chiral catalyst used in their study was derived from 1,1'-bis(diphenylphosphino)ferrocene anchored to the inner wall, in single-site fashion, of a mesoporous silica, similar to that depicted in Figure 7.11. There is little doubt that the overall performance of this one-step process — both the degree of conversion and the *ee* — could be further improved by appropriate fine-tuning of the experimental conditions, including the use of diamine-based chiral ligands.

7.5 Epilogue

Three key facts may be extracted from the topics discussed in this chapter:

- Although encouraging progress has been made to prepare chiral zeolites and AlPOs, no indisputable proof has yet appeared for the preparation of chiral organic products using such (chiral) solids as catalysts. The main difficulty in the use of these SSHCs of a microcrystalline kind is the production of homochiral solid catalysts.

- MOF single-site catalysts, largely because of the relative ease of preparing them in homochiral form, have already proved successful in generating chiral products (such as alcohols from ketones). Such successes have so far been restricted to laboratory-scale operations.

- Chiral organometallic compounds, confined within mesoporous silicas, are very efficient enantioselective SSH catalysts for a variety of reactions, encompassing selective hydrogenation and oxidation. Although used mainly, to date, at the laboratory scale, at least one commercial process uses this type of SSHC for the production of asymmetric ketoester by enantioselective hydrogenation.

Industrially, enantioselective catalysis with metal complexes, is, at present, mainly carried out in homogeneous solution, notwithstanding the practical difficulties associated with separations and recyclability.[59,61,62]

References

1. R. M. Barrer and P. J. Denny. Hydrothermal chemistry of the silicates. Part IX. Nitrogenous aluminosilicates, *J. Chem. Soc.*, 971 (1961).
2. J. M. Newsam, M. M. J. Treacy, W. T. Koetsier and C. B. de Gruyter. Structural characterization of zeolite beta, *Proc. R. Soc. Lond. A*, **420**, 375 (1988).
3. L. Q. Tang, L. Shi, C. Bonneau, J.-L. Sun, H. J. Yue, A. Ojuva, B. L. Lee, M. Kritikos, R. G. Bell, Z. Bacsik, J. Mink and X. D. Zou. A zeolite family with chiral and achiral structures built from the same building layer, *Nat. Mater.*, **7**, 381 (2008).
4. K. D. M. Harris and J. M. Thomas. Selected thoughts on chiral crystals, chiral surfaces and asymmetric heterogeneous catalysts, *Chem. Cat. Chem.*, **1**, 223 (2009).
5. R. E. Morris and X. Bu. Induction of chiral porous solids containing only achiral building blocks, *Nat. Chem.*, **2**, 353 (2010).
6. S. Ramdas, J. M. Thomas, M. E. Jordan and C. E. Eckhardt. Enantiomeric intergrowths in hexahelicenes, *J. Phys. Chem.*, **85**, 2421 (1981).

7. The crystal structure of hexahelicene has the chiral space group $P2_12_12_1$. However, it is found that single crystals grown from racemic solution are virtually non-chiral, possessing *ee* values of only a few percent. The reason for this seemingly puzzling phenomenon is that, as shown in Ramdas *et al.*,[6] individual crystals contain lamellar intergrowths of approximately equal amounts of the two enantiomorphous forms of the chiral structure.

8. Lord Kelvin. *Baltimore Lectures (1884) on Molecular Dynamics and the Wave Theory of Light,* Clay and Sons, London (1904).

9. B. Kahr, Y. Bing, W. Kaminsky and D. Viterbo. Turinese stereochemistry: Eligio Perucca's enantioselectivity and Primo Levi's asymmetry, *Angew. Chem. Int. Ed.,* **48**, 2 (2009).

10. W. T. A. Harrison, T. E. Gier, G. D. Stucky, R. W. Broach and R. A. Bedard. $NaZnPO_4 \cdot H_2O$: an open-framework sodium zincophosphate with a new chiral tetrahedral framework topology, *Chem. Mater.,* **8**, 145 (1996).

11. X. W. Song, Y. Li, L. Gan, Z. Wang, J. Yu and R. Xu. Heteroatom-stabilized chiral framework of aluminophosphate molecular sieves, *Angew. Chem. Int. Ed.,* **48**, 314 (2009).

12. J. H. Yu and R. R. Xu. Chiral zeolitic materials: structural insights and synthetic challenges, *J. Mater. Chem.,* **18**, 4021 (2008).

13. J. L. Sun, C. Bonneau, A. Cantin, A. Corma, M. J. Diaz-Cabanas, M. Moliner, D. L. Zhang, M. R. Li and X. D. Zhou. The ITQ-37 mesoporous chiral zeolite, *Nature,* **458**, 1154 (2009).

14. R. E. Morris. Some difficult challenges for the synthesis of nanoporous materials, *Top. Catal.,* **53**, 1291 (2010).

15. E. R. Parnham and R. E. Morris. Ionothermal synthesis of zeolites, metal-organic frameworks and inorganic-organic hybrids, *Acc. Chem. Res.,* **40**, 1005 (2007).

16. M. E. Davis. Reflections on routes to enantioselective solid catalysts, *Top. Catal.,* **25**, 1 (2003).

17. K. A. Avery, R. Mann, M. Norton and D. J. Willock. Computer simulation of structural aspects of enantioselective heterogeneous catalysis and the prospects of direct calculation of selectivity, *Top. Catal.,* **25**, 89 (2003).

18. G. J. Hutchings. New approaches to rate enhancement in heterogeneous catalysts, *Chem. Commun.,* 301 (1999).

19. G. D. Wu and W. Lin. Heterogeneous asymmetric catalysis with homo-chiral metal-organic frameworks: network-structure-dependent catalytic activity, *Angew. Chem. Int. Ed.*, **46**, 1075 (2007).

20. (a) R. Vaidhyanathan, D. Bradshaw, J. N. Rebilly, J. P. Barrio, J. A. Gould, N. G. Berry and M. J. Rosseinsky. A family of nanoporous materials based on an amino acid backbone, *Angew. Chem. Int. Ed*, **45**, 6495 (2006).

    (b) A. Kuschel and S. Polarz. Effects of primary and secondary surface groups in enantioselective catalysis using nanoporous materials with chiral walls, *J. Am. Chem. Soc.*, **132**, 6558 (2010).

21. M. J. Ingleson, J. P. Barrio, J. Bacsa, C. Dickinson, H. Park and M. J. Rosseinsky. Generation of a solid Brønsted acid site in a chiral framework, *Chem. Commun.*, 1287 (2008).

22. K. P. Lillerud, K. Olsbye and M. Tilset. Designing heterogeneous catalysts by incorporating enzyme-like functionalities into MOFs, *Top. Catal.*, **53**, 859 (2010).

23. (a) L. Ma, J. M. Falkowski, C. Abney and W. Lin. A series of isoreticular chiral metal-organic frameworks as a tunable platform for asymmetric catalysis, *Nat. Chem.*, **2**, 838 (2010).

    (b) See special issue, Catalysis by MOFs: Quo Vadis?, *Phys. Chem. Chem. Phys.*, **13**, 6373 (2011).

    (c) V. Colombo, S. Galli, H. J. Choi, G. D. Han, A. Maspero, G. Palmisano, N. Masciocchi and J. R. Long. High thermal and chemical stability in pyrazolate-bridged metal-organic frameworks with exposed metal sites, *Chem. Sci.*, **2**, 1311 (2011).

24. A. Baiker. Chiral catalysts on solids, *Curr. Opin. Solid State Mater. Sci.*, **3**, 86 (1998).

25. S. B. Ogunwumi and T. Bein. Intrazeolite assembly of a chiral Mn salen epoxidation catalyst, *Chem. Commun.*, 901 (1997).

26. A. Corma, A. Fuerte, M. Iglesias and F. Sanchez. Preparation of new chiral dioxomolybdenum complexes heterogenized on modified USY-zeolites efficient catalysts for selective epoxidation of allylic alcohols, *J. Mol. Catal.*, **107**, 225 (1996).

27. C. E. Song, J. W. Yang, H. J. Ha and S. Lee. Efficient and practical polymeric catalysts for heterogeneous asymmetric dihydroxylation of olefins, *Tetrahedron Asymmetry*, **7**, 645. (Quoted as Ref. 52 in Ref. 24.)

28. S. Shimazu, K. Ro, T. Sento, N. Ichikuni and T. Heinatsu. Asymmetric hydrogenation of α,β-unsaturated carboxylic acid esters by rhodium(I)-phosphine complexes supported on smectites, *J. Mol. Catal.*, **107**, 297 (1996).

29. J. M. Thomas. Tales of tortured ecstasy: probing the secrets of solid catalysts, *Faraday Disc.*, **100**, C9 (1995).

30. J. M. Thomas, T. Maschmeyer, B. F. G. Johnson and D. S. Shephard. Constrained chiral catalysts, *J. Mol. Catal.*, **141**, 139 (1999).

31. (a) B. M. Trost and E. Keinan. Steric steering with supported palladium catalysts, *J. Am. Chem. Soc.*, **100**, 7779 (1978).

    (b) J. Tsuji. *Palladium Reagents and Catalysis: Innovation in Organic Synthesis*, Wiley, New York (1995).

32. B. F. G. Johnson, S. A. Raynor, D. S. Shephard, T. Maschmeyer, J. M. Thomas, G. Sankar, S. T. Bromley, R. D. Oldroyd, L. F. Gladden and M. D. Mantle. Superior performance of a chiral catalyst confined within mesoporous silica, *Chem. Commun.*, 1167 (1999).

33. Silsesquioxanes, general formula $RSiO_{3/2}$ where R = H, alkyl, aryl, etc. These have cage-like structures. Not only are they models for a silica surface, they may also be regarded as soluble chunks of silica to which may be attached a heteroatom M, thus yielding $Si_7MO_{12}R_8$.

34. S. A. Raynor, J. M. Thomas, R. Raja, B. F. G. Johnson, R. G. Bell and M. D. Mantle. A one-step, enantioselective reduction of ethyl nicotinate to ethyl nipecotinate using a constrained, chiral, heterogeneous catalyst, *Chem. Commun.*, 1925 (2000).

35. R. Raja, J. M. Thomas, M. D. Jones, B. F. G. Johnson and D. E. W. Vaughan. Constraining asymmetric organometallic catalysts within mesoporous supports boosts their enantioselectivity, *J. Am. Chem. Soc.*, **125**, 14982 (2003).

36. M. D. Jones, R. Raja, J. M. Thomas, B. F. G. Johnson, D. W. Lewis, J. Rouzard and K. D. M. Harris. Enhancing the enantioselectivity of novel homogeneous organometallic hydrogenation catalysts, *Angew. Chem. Int. Ed.*, **42**, 4326 (2003).

37. J. Rouzard, M. D. Jones, R. Raja, B. F. G. Johnson and J. M. Thomas. Potent new heterogeneous asymmetric catalysts, *Helv. Chim. Acta*, **86**, 1753 (2003).

38. C. Li, H. D. Zhang, D. M. Jiang and Q. H. Yang. Chiral catalysis in nanopores of mesoporous materials, *Chem. Commun.*, 547 (2007).

39. K. Malek, A. P. J. Jansen, C. Li and R. A. van Santen. Enantioselectivity of immobilized Mn-salen complexes: a comparative study, *J. Catal.*, **246**, 127 (2007).

40. C. E. Song and S. G. Lee. Supported chiral catalysts on inorganic materials, *Chem. Rev.*, **102**, 3495 (2002).

41. C. E. Song, D. H. Kim and D. S. Choi. Chiral organometallic catalysts in confined nanospaces: significantly enhanced enantioselectivity and stability, *Eur. J. Inorg. Chem.*, 2927 (2006).

42. J. M. Thomas and R. Raja. Designing catalysts for clean technology, green chemistry and sustainable development, *Annu. Rev. Mater. Sci.*, **35**, 315 (2008).

43. J. M. Thomas and R. Raja. Exploiting nanospace for asymmetric catalysis: confinement of immobilized, single-site chiral catalysts enhances enantioselectivity, *Acc. Chem. Res.*, **41**, 708 (2008).

44. F. M. de Rege, D. K. Morita, K. C. Ott, W. Tumas and R. D. Broene. Non-covalent immobilization of homogeneous cationic rhodium-phosphine catalysts on silica surfaces, *Chem. Commun.*, 1797 (2000).

45. These are supplied by the W. R. Grace Co., USA. They have greater attrition resistance (see Chapter 4, Section 4) than the rather fragile MCM family of silicas which are, in any case, more expensive and require thorough prior elimination of the organic template before use.

46. P. Piaggo, P. McMorn, D. Murphy, D. Bethell, P. C. B. Page, F. E. Hancock, C. Sly, O. J. Kerton and G. J. Hutchings. Enantioselective epoxidation of (Z)-stilbene using a chiral Mn(III)-salen complex: effect of immobilisation on MCM-41 on product selectivity, *J. Chem. Soc. Perkin Trans.*, 2008 (2000).

47. H. Zhang and C. Li. Asymmetric epoxidation of 6-cyano-2, 2-dimethylchromene on Mn(salen) catalyst immobilized in mesoporous materials. *Tetrahedron*, **62**, 6640 (2006).

48. G. Gbeny, A. Zsigmond and K. J. Balkus. Enantioselective epoxidations catalyzed by MCM-22 encapsulated Jacobsen's catalyst, *Catal. Lett.*, **74**, 77 (2001).

49. A. Corma, H. Garcia, A. Moussaif, M. Sabater, R. Zniber and R. Redouane. Chiral copper(II) bisaxazoline covalently anchored to silica and mesoporous MCM-41 as a heterogeneous catalyst for the enantio-selective Friedel–Crafts hydroxyalkylation, *Chem. Commun.*, 1058 (2002).

50. Y. Tashiro, T. Nagashima, S. Aoki and R. Nishizawa. Process for preparing optically active amino acid or mandelic acid, U.S. Patent No. 4, 224, 239 (1980).

51. T. Endo and K. Tamura. Process for producing R(-)-mandelic acid and derivatives thereof, U.S. Patent No. 5, 223, 416 (1993).

52. E. Haslam. Recent developments in methods for the esterification and protection of the carboxyl group, *Tetrahedron*, **36**, 2409 (1980).

53. G. D. Yadav and R. D. Bhagat. Synthesis of methyl phenyl glyoxalate via clean oxidation of methyl mandelate over a nanocatalyst based on heteropolyacid supported on clay, *Org. Process. Res. Dev.*, **8**, 879 (2004).

54. Y. G. Zhou. Asymmetric hydrogenation of heteroaromatic compounds, *Acc. Chem. Res.*, **40**, 1357 (2007).

55. F. Fache, E. Schulz, M. L. Tommasino and M. Lemaire. Nitrogen-containing ligands for asymmetric homogeneous and heterogeneous catalysis, *Chem. Rev.*, **100**, 2159 (2000).

56. J. M. Thomas, B. F. G. Johnson, R. Raja and M. Jones. Process for asymmetrically hydrogenating ketocarboxylic esters, U.S. Patent No. 0220165 AL (04/11/2004).

57. J. M. Thomas, B. F. G. Johnson, R. Raja and M. Jones. Verfahren zur Reduktion von Ketocarbonsäuerestern. E.P. Patent No. 1469006, A2 (20/10/2004).

58. D. R. Penman, G. O. Osborne, S. P. Worner, R. B. Chapman and G. F. McLaren. Ethyl nicotinate: a chemical attractant for *Thrips obscuratus* in stonefruit in New Zealand, *J. Chem. Ecology*, **8**, 1299 (1982).

59. H. U. Blaser, H. Hönig, M. Studer and C. Wedemeyer-Exl. Enantioselective synthesis of ethyl nipecotinate using cinchona modified heterogeneous catalysts, *J. Mol. Catal. A*, **139**, 253 (1999).

60. S. A. Raynor, J. M. Thomas, R. Raja, B. F. G. Johnson, R. G. Bell and M. D. Mantle. A one-step, enantioselective reduction of ethyl nicotinate to ethyl nipecotinate using a constrained, chiral, heterogeneous catalyst, *Chem. Commun.*, 1925 (2000).

61. R. Noyori. Asymmetric catalysis: science and opportunities, *Angew. Chem. Int. Ed.*, **41**, 2008 (2002).

62. R. Noyori, M. Kitamura and T. Ohkuma. Toward efficient asymmetric hydrogenation: architectural and functional engineering of chiral molecular catalysts, *PNAS*, **101**, 5356 (2004).

CHAPTER 8

MULTINUCLEAR, BIMETALLIC NANOCLUSTER CATALYSTS

8.1 Definitions: Nanoclusters are Distinct from Nanoparticles

We first need to distinguish nanoclusters from nanoparticles. In nanocluster catalysts the total number of atoms is very small, ranging from three or four to no more than about 20. In nanoparticles, on the other hand, where one is concerned with particle sizes 5 to 15 nm in diameter[1-3] the number of atoms in the nanoparticle falls in the range from 10^4 to 10^{10}. This difference alone leads to other important distinctions. The first of these concerns the so-called dispersion, D, which is the fraction of surface-situated atoms in a given particle cluster. In an eight-atom cluster D is 1.0; in a 125-atom cluster D is 0.78; and with 3,000 atoms D drops to 0.4. The second difference devolves upon crystallographic factors (or their lack). Whereas it makes perfect sense to identify distinct planes or facets (like the (111) and (100) faces) in the kind of Pd nanoparticles studied by Freund and co-workers,[1] nanoclusters, owing to their minuteness, do not possess such defined faces.

Because of their essentially molecular character, the nanocluster catalysts considered here (typically Ru_6Pd_6) have no continuous band of energies; instead they possess discrete energy levels, as indicated by their computed density of states (DOS) plots shown in Figure 8.1. It is to be noted that the energies of the lowest unoccupied molecular orbital (LUMO) and highest occupied molecular orbital (HOMO) for a Pt_3 cluster are substantially higher than the Fermi energy of a nanoparticle consisting of 309 Pt atoms. The

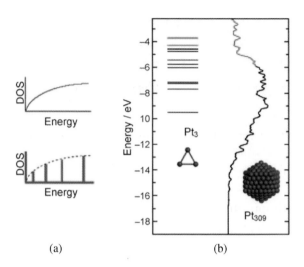

Figure 8.1 (a) Schematic illustration of the density of states (DOS) of the electronic energy levels in a typical metallic nanoparticle (top) and in a typical nanocluster (bottom); (b) Computed energy levels in a Pt_3 nanocluster (left) and a Pt_{309} nanoparticle (right). Occupied energy levels shown in black.

positions in energy space of the LUMOs and HOMOs therefore enable electron capture or electron donation (respectively) to or from the nanocluster and a reacting species to be facilitated in comparison with a nanoparticle of the same chemical elements. In other words, surface reactions should proceed more smoothly at the nanocluster. As a consequence, we expect, and we do indeed find — as we show in this chapter — that the catalytic activities of small nanoclusters are very high. Moreover, rapid molecular transformations are expected when these nanoclusters — a reliable synonym is "molecular metal" — are composed of atoms of metal M known to form single or multiple bonds of intermediate strength, as required for catalysis involving the constituent elements of organic compounds (C, H, N, O, S...), where M may be Pt, Pd, Ag, Au, Cu or Ru.

In their early work on the bimetallic nanocatalysts to be summarized in this chapter, the author and his colleagues did not emphasize adequately the distinctions between nanoclusters and nanoparticles. (Often the term "nanoparticle" was used in place of "nanoclusters").

8.1.1 *Bimetallic nanoclusters and bimetallic nanoparticles are not alloys*

It was Sinfelt[4] in the early 1980s who first used the term "bimetallic" to describe the kind of two-component species that effect the catalytic performance that is of interest to us here. In his own words: "... since the ability to form bulk alloys was not a necessary condition for a system to be of interest as a catalyst, it was decided not to use the term alloy in referring to bimetallic catalysts in general. In particular, bimetallic clusters refer to bimetallic entities which are highly dispersed on the surface carrier."

These definitions hold here. The key difference, however, between our type of bimetallic clusters[5] and the nanoparticles described by Sinfelt is this: in Sinfelt's bimetallic nanoparticles, such is their size that surface atoms constitute only a relatively small fraction (less than 10%) of the total number of metal atoms in the catalysts, whereas in ours there is a very high fraction (greater than 90% and often 100%). This is because our nanoclusters are much smaller, the diameter of the nanoclusters often being less than 1 nm.

In our nanoclusters, because of the way they are introduced, the individual entities are distributed and anchored in a spatially uniform manner along the inner walls of high-area (500 to 900 $m^2 g^{-1}$) nanoporous silica supports, the sharply defined pore diameter of which is controllable within the range of 3 to 30 nm. Because the bimetallic nanocatalysts are firmly anchored through the oxophilic nature of one of the constituents to the walls of these (relatively large) pores, their tendency to sinter and coalesce (with consequential loss of active area) is minimized; and, in addition, there is essentially free diffusional access of reactants to, and egress of products from, the dispersed nanoclusters, thereby facilitating turnover of quite bulky organic molecules.[6,7] The essential architectural features of these novel nanocatalysts are depicted in Figures 8.2 and 8.3.

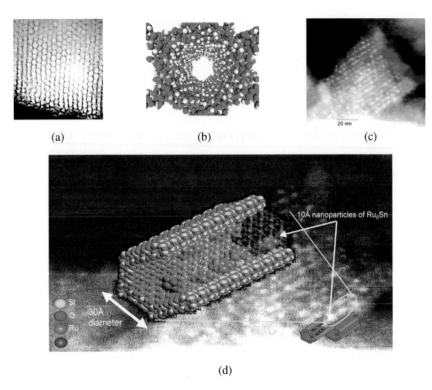

(a)   (b)   (c)

(d)

Figure 8.2 (a) High-resolution micrograph of a typical hexagonal array of nanopores in silica (100 Å diameter). (b) Computer graphic representation of interior of a single pore of mesoporous silica, showing pendant silanol groups. (c) High-angle annular dark-field (scanning electron transmission) micrograph showing the distribution of Ru_6Sn nanoclusters within the pores of the siliceous host. (d) Computer graphic illustration of the Ru_6Sn nanoclusters superimposed on an enlargement of the electron micrograph shown in (c). (After Thomas *et al.*[5])

8.2 The Merits of Studying Bimetallic Nanocluster Catalysts

When my colleagues and I embarked on the study of these bimetallic nanocatalysts, we were motivated by several perceived advantages:

- First, we were already aware (from FTIR, NMR and other evidence) that gentle thermolysis of metal cluster (and especially mixed-metal cluster) carbonylates (typically $[Ru_6Pd_6(CO)_{24}]^{2-}$),

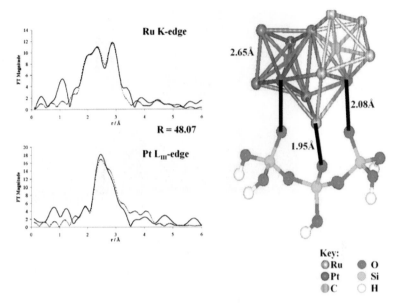

Figure 8.3 Illustration of how the Ru K- and Pt LIII-edges in X-ray absorption spectra yield the structure of the $Ru_{10}Pt_2C_2$ nanocluster catalyst anchored to mesoporous silica. (After Thomas *et al.*[32])

when attached to oxide surfaces rich in pendant hydroxyl groups, could yield individual (bare, like Ru_6Pd_6) decarbonylated metallic or bimetallic nanoclusters strewn across the oxide surfaces to which the carbonylate species had originally been attached. (We soon confirmed this fact in a series of combined[8] *in situ* XAFS and FTIR studies; the results obtained with a typical mixed-metal parent carbonylate $[Ru_{12}C_2(CO)_{32}Cu_4Cl_2]^{2-}$ anion are shown in Figure 8.4.)

- Second, we were also aware that, potentially, a vast number[9] of precursor metal and mixed-metal carbonylates, of well-defined structure and stoichiometry, were available for the production of a wide range of novel nanocatalysts. A small selection of such precursor materials is enumerated in Table 8.1.

- Third, the great advantage in preparing bimetallic nanocatalysts from such precursors is that it guarantees — as many subsequent elemental fingerprinting and electron microscopic

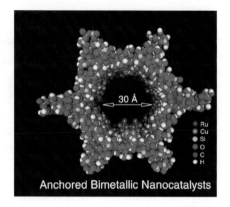

Figure 8.4 Representation of the manner in which a $Ru_{12}Cu_4$ nanocluster is anchored inside the pores of mesoporous silica. (After Thomas and Bell, unpublished work.)

Table 8.1 Typical parent anionic carbonylates from which naked nanoclusters (*ca* 0.1–0.2 nm diameter) are generated.[5]

$[Ru_6Pd_6(CO)_{24}]^{2-}$	$[Pt_3Cu(CO)_3(PPh_3)]^-$
$[Ru_6C(CO)_{16}SnCl_3]^{2-}$	$[Ru_5PtC(CO)_{15}]^{2-}$
$[Ru_{12}C_2(CO)_{32}Cu_4Cl_2]^{2-}$	$[Ru_{10}Pt_2C_2(CO)_{28}]^{2-}$

imaging studies have repeatedly confirmed[10,11] — the integrity (and stoichiometry) of the nanocluster catalyst, an end that is difficult to achieve using the incipient wetness or co-precipitation method of preparation.[4] Figure 8.5 illustrates[12] this feature well.

- Fourth, it had already been established in other contexts[13] — involving high-temperature, hydrocarbon-reforming catalysts — that there was special merit in using bimetallic, rather than monometallic, nanoclusters in that their catalytic performance is generally far superior to that of single nanoclusters alone. For example, we quickly discovered[14] that nanocatalysts of Pd and Ru alone were each far less active and less selective as hydrogenation catalysts than their Ru_6Pd_6 bimetallic counterparts (see Table 8.2). We also observed that bimetallic catalysts were more resistant to

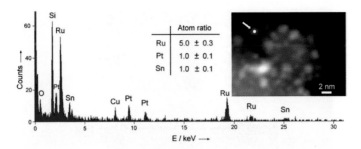

Figure 8.5 A plot showing the uniformity of the composition of several nanoclusters of Ru_5PtSn (anchored onto a mesoporous silica) obtained by electron-induced X-ray emission, with the diameter of the probing beam *ca* 10 Å. (After Thomas *et al.*[18])

Table 8.2 Hydrogenation of 1-hexane with Ru_6Pd_6 and Ru_6 nanoclusters and Pd nanoparticles supported on mesoporous silica.[14]

Reactant (g)	Catalyst	Reaction Time (h)	Conversion (%)	TOF (h⁻¹)	Products (mol %) A	B	C
1-Hexene (50 g)	Ru_6Pd_6/SiO_2	4	99	4954	68	22	9
1-Hexene (50 g)	Ru_6/SiO_2	4	13	325	14	42	45
		24	19	277	10	36	53
1-Hexene (50 g)	Pd/SiO_2 (nanoparticles)	4	6	250	6	45	48
		24	14	196	5	33	63
	None	24	7	—	—	32	67

Reaction conditions: catalyst, 20 mg; temperature, 373 K; starting H_2 pressure, 20 bar; no solvent.
A = *n*-hexane; B = *cis*-2-hexene; C = *trans*-2-hexene.

sulfur poisoning (in their role as facilitators of partial hydrogenation of naphthalene) than the separate components alone (as nanocatalysts).[15]

• Fifth, special preparative *in situ* X-ray absorption spectroscopic (and other) techniques[16] allied to powerful *ex situ* high-resolution

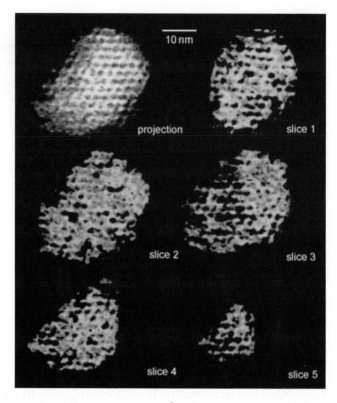

Figure 8.6 An axial projection of a 300 Å thick specimen and five successive 30 Å thick slices through a tomogram of silica-supported $Ru_{10}Pt_2C_2$ nanocluster catalysts (red). (After Thomas *et al.*[18])

electron microscopic ones[17] were available for the determination of the detailed atomic structure of the bare, supported bimetallic nanocatalysts. We were also aware, initially, of the great insights that electron tomography might provide: these have been amply fulfilled[11,17] (see Figure 8.6). The particular advantage of using Rutherford scattering — i.e. using high-angle, annular dark-field (HAADF) scanning electron microscopy — is that it is an ideal means of locating (high Z) heavy atoms where supported on a light (e.g. SiO_2) background[19] (see Figure 8.2(c)).

- The sixth point to emphasize concerns the surface cleanliness of the bimetallic cluster catalysts (prepared from mixed-metal

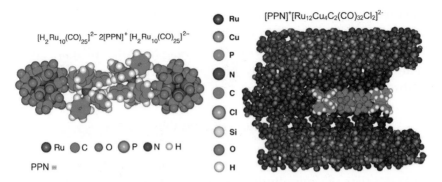

Figure 8.7 A metal carbonylate (anion) along with an organic cation (PPN = bis(triphenylphosphine)iminium) is sequestered into the nanopores of silica. On gentle heating in O_2 only the naked clusters remain, as evidenced by XAFS and FTIR measurements. (After Thomas *et al.*[68])

carbonylates). The so-called nanoalloy and nanoparticle (bimetallic) catalysts that many workers have studied are generally prepared by the incipient wetness method that relies on sequential addition of separate solutions such as $HAuCl_4$ and $PdCl_2$ (to form Au-Pd nanoparticles). Moreover, many workers, see, for example, Rioux *et al.*,[20] utilize their nanocatalysts in passivated forms, which means that residual surfactant species such as poly(vinylpyrrolidone) (PVP) or poly(diallyldimethylammonium chloride) (PDDA), the molecular weights of which may be as large as 450,000 Da, cover part of the nanocatalyst surface.

Finally, the method of preparing the bimetallic nanocatalysts, by sequestering them in mesopores as bulky organometallic precursors (as illustrated in Figure 8.7), ensures that the denuded nanoclusters are spatially well separated from one another.

8.3 Why Focus on Bimetallic Catalysts Based on Platinum Group Metals (PGMs)?

There are many bimetallic catalysts of great industrial and other relevance that contain no platinum group metals (PGMs) as constituent — an outstanding example is Raney NiSn, which is particularly important[21] in the aqueous-phase reforming (APR) of derivatives from biomass (e.g. sorbitol, glycerol, ethylene glycol) that, in an environmentally benign fashion, convert oxygenated hydrocarbons into hydrogen:

$$C_2H_6O_2 \ (l) + 2H_2O \ (l) \rightleftarrows 2CO_2 \ (g) + 5H_2 \ (g).$$

However, the large majority of industrial catalysts are composed of PGMs. For reasons alluded to earlier, all these metals (M) readily form chemical bonds that are of intermediate strength with elements such as C, N, H, O and S; and this is one of the key desiderata for an effective catalyst.[22,23] It is not surprising, therefore, that, both for academic and commercial reasons, bimetallic nanocatalysts with PGMs as constituents are especially popular. Not only does one observe the high catalytic activity, selectivity and durability of such catalysts, they also exhibit very important kinds of bifunctionality: they may readily facilitate hydrogenation (and hence, by the principle of microscopic reversibility dehydrogenation) of hydrocarbons.[24] Many of them catalyse isomerizations and possess admirable reforming attributes.

Nowadays, however, PGMs are so extensively used for a wide variety of catalytic hydrogenations, isomerizations and oxidations that there is an exigent need to seek effective, cheaper and more readily available substitutes. Whereas there are few indications, at present, that total substitution (e.g. for Pt in fuel cell assemblies and for Rh, Pd and Pt in auto exhausts) will shortly become a reality, what is clear is the need to use bimetallic catalytic species in which a second constituent is bound to a member (or members) of the PGMs. Previously, Siani *et al.*[25] described how FePt nanocluster catalysts function effectively in the so-called PROX (preferential oxidation) process for the selective oxidation of carbon monoxide in the presence of H_2 — a reaction of considerable importance in both ammonia synthesis and the Fischer–Tropsch reaction, where, in each case, traces of CO poison the active catalyst:

$$2CO + H_2 + O_2 \rightarrow 2CO_2 + H_2.$$

In association with R. D. Adams and his team, my colleagues and I have extended[26,27] my earlier work (with B. F. G. Johnson and R. Raja) in using Ru-Sn bimetallic nanocatalysts, as well as using some trimetallic ones, in which Sn is incorporated into a cluster of two PGMs. A deep understanding of why the catalytic performance of bimetallic (and trimetallic) nanocluster catalysts is so outstanding is still awaited, but some preliminary thoughts, based on recent theoretical investigations,[28,29] are considered below. For heuristic and other purposes we proceed to regard the individual multinuclear bimetallic cluster as a single site for the locus of the catalytic reaction in question. Our results are quoted with reference to the number of clusters in a sample as being the number of active sites present. Clearly, if only a few or a small ensemble within the nanocluster is the effective (multinuclear) active site, then our quoted results, given in Table 8.3, constitute lower estimates of activity. We first summarize a number of significant conversions,

Table 8.3 A selection of highly active and selective bimetallic cluster catalysts for the single-step hydrogenation of some key organic compounds.[5,28,66]

Catalyst	Reaction	Solvent	TOF (h^{-1})	Uses of product
Ru_6Pd_6/SiO_2	Cyclododecatriene	—	5350	Laurolactam,
Ru_6Sn/SiO_2	(CDT) to		1940	copolyamides,
	cyclododecene			nylons
	(CDE)			
Ru_6Pd_6/SiO_2	Cyclooctadiene to	—	2010	Polymer
Ru_6Sn/SiO_2	the monoene		1980	intermediates,
				ketones and
				polyesters
Ru_5Pt/SiO_2	Benzene to	—	2625	Starting material
$Ru_{10}Pt_2/SiO_2$	cyclohexene		1790	in production of
Ru_6Pd_6/SiO_2			3210	K-A oil (a mixture
				of cyclohexanol
				and the ore)
Ru_5Pt/SiO_2	Phenol to K-A oil	Hexane	450	Precursor for
Ru_6Pd_6/SiO_2			510	production of
				ε-caprolactam
				and nylon

particularly selective hydrogenations, that my colleagues and I have investigated.

8.4 Specific Examples of High-performance Bimetallic Nanocluster Catalysts for Selective Hydrogenations under Benign Conditions

Figure 8.8 and Table 8.3 summarize the results obtained in a variety of hydrogenation processes utilizing a number of different catalysts. The results fall into two categories: those that operate under solvent-free conditions, and those that require a benign solvent.

Many of the products shown in Figure 8.8 are of great industrial importance. For example, the principal product of selective hydro-genation of 1,5,9-cyclododecatriene (CDT), to cyclododecene (CDE) — see Scheme 8.1 — is the source of nylon 12, nylon 612, polyesters, copolyamides and materials for coating applications.

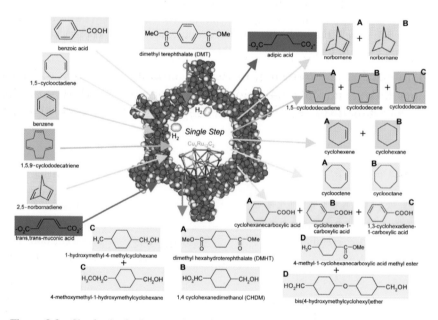

Figure 8.8 Single-site hydrogenation of some key organic compounds using highly active and selective, anchored bimetallic nanocluster catalysts ($Ru_{12}Cu_4C_2$ in this instance).

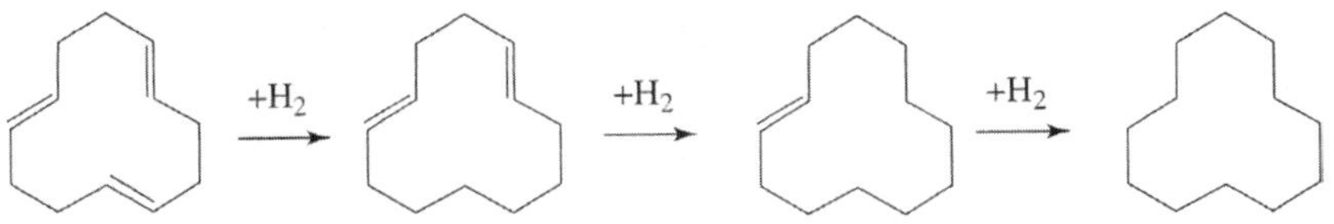

Scheme 8.1 The stepwise reduction of cyclododecatriene (CDT) to cyclododecane (CDA).

A wide variety of homogeneous and heterogeneous catalysts such as Raney Ni, Pd, Pt, Co and mixed-transition-metal complexes has been used previously for the hydrogenations shown in Figure 8.8 and Table 8.3. But all these reactions entailed the use of organic solvents (such as *n*-heptane and benzonitrile), and some required utilization of hydrogen donors (such as 9,10-dihydroanthracene) — often at temperatures in excess of 300 °C — to achieve the desired selectivities.

The exceptional synergistic catalytic performance of the anchored $\{Ru_{12}Cu_4C_2\}$ nanocluster (as well as its atomic structure, determined as described later) is shown in Figure 8.9 — see also Section 8.6. Here we demonstrate the supreme advantages of this bimetallic nanocatalyst in producing, in cascade fashion, the important polymer linker molecule 1,4-cyclohexanedimethanol (CHDM) from its parent dimethyl terephthalate (DMT) *via* the intermediate dimethyl hexahydroterephthalate (DMHT). At present, the favoured industrial method of producing CHDM (a versatile linker molecule in polymer chemistry known to be superior to ethylene glycol) involves two distinct and demanding processes (Figure 8.9(a)). DMT is first partially hydrogenated to DMHT using a highly exothermic process at elevated pressure and temperature over a Pd catalyst, and it, in turn, is converted over a copper chromite catalyst at 493 K using 4 MPa of H_2[30] to the desired CHDM. The synergy in the anchored bimetallic nanocluster catalyst affects both these steps and generates the CHDM directly from DMT in one step at relatively mild conditions. Significantly, taken in isolation, the separate components of our bimetallic nanocluster, Cu and Ru, are both inferior with respect to the degree of conversion and selectivity — see Figure 8.9(c).

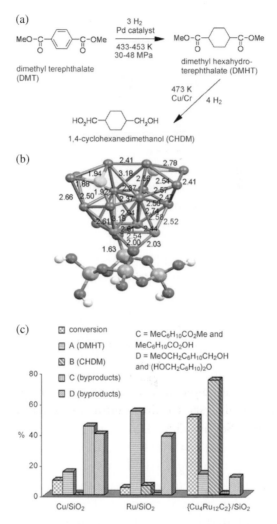

Figure 8.9 $Ru_{12}Cu_4C_2$ anchored nanoclusters shown in (b) catalyse in cascade fashion the two reactions shown in (a), with good selectivity at 353 K and 20 bar H2 (c). (After Thomas and Raja, *Top. Catal.*, **53**, 848 (2010).)

8.4.1 *Bimetallic nanocluster catalysts for ammoxidation*

It is interesting to note that the $Ru_{12}Cu_4C_2$ nanocluster is also an effective ammoxidation catalyst,[31] as seen in Figure 8.10, where it is shown that good turnover frequencies and conversion are obtained

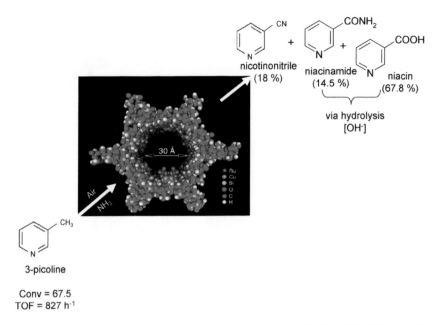

Figure 8.10 The selective ammoxidation of 3-picoline (403 K; 2 MPa NH₃; 4 MPa air) to yield vitamin B3 (niacin). (After Thomas *et al.*[31])

in the conversion of 3-picoline to nicotinonitrile, which is readily converted by hydrolysis to the important vitamin B_3 (niacin), described fully in Section 5.2.

8.4.2 *Bimetallic nanocluster catalysts for the (sustainable) synthesis of adipic acid*

It has been found[5,6,32] that bimetallic cluster catalysts, such as $Ru_{12}Pt_2C_2$, display high performance in converting muconic acid (readily derived by biocatalysis from glucose) to adipic acid (see Figure 8.11). This is a one-step, solvent-free process that takes place at low temperatures. This novel route for producing adipic acid avoids the use of concentrated nitric acid as the oxidant with the consequential generation of large volumes of greenhouse gases such as N_2O. It is seen from Figure 8.11 that the nanocluster $Ru_{10}Pt_2C_2$ (shown top right) is exceptional in its selectivity compared to other

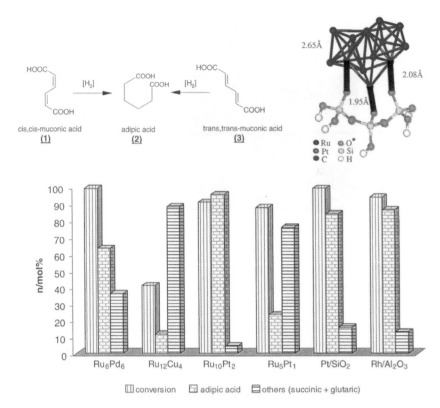

Figure 8.11   Conversion of muconic acid to adipic acid using $Ru_{10}Pt_2C_2$ cluster nanocatalysts anchored to mesoporous silica. (After Thomas *et al.*[32])

bimetallic nanoclusters (such as Ru_6Pd_6 and $Ru_{12}Cu_4C_2$). It is also superior to industrially used monometallic supported nanoclusters such as Pt and Rh. This augurs well, as noted in an Editorial in the magazine *Science*,[32] for the future use of high-area, thermally stable, nanocluster catalysts in a wide range of other hydrogenations that may be effected to yield desirable chemical products from plant crop sources.

The *Science* article ("Highlights of the Recent Literature") is called "Renewable Nylons" (May 2003) and reads as follows:

> Many efforts aimed at improving the environmental friendliness of
> chemical processes tend to focus on minimizing unwanted side

products or solvent use, or on replacing a fossil fuel-based material with a different one based on a renewable feedstock. Thomas *et al.*[32] now report on a catalytic method for producing a key intermediate that allows the same product, in this case nylon, to be made from renewable sources. Adipic acid is widely used in the production of nylon and is usually produced from fossil fuel-derived benzene. Thomas *et al.* used muconic acid, which can be derived from glucose, as the starting material. Their catalysts consisted of bimetallic clusters anchored in the pores of mesoporous silica. A Ru_2Pt_{10} composition was more selective than several other bimetallic clusters and was also superior to monometallic Pt and Rh catalysts. Bimetallic catalysts of this kind may be useful for other environmentally friendly hydrogenation reactions.

8.5 Bimetallic and Trimetallic Nanocluster Catalysts Containing Tin: The Experimental Facts

It has long been known that the catalytic activity of PGMs (like that of some transition metals[21]) is strongly modified by the presence of appropriate amounts of tin.[12,33–35] In addition to its catalytic influence, there is also evidence[6] that tin helps to anchor metallic nanocatalysts onto nanoporous supports. Ph_3SnH is an excellent reagent for introducing variable numbers of phenyltin-containing ligands into polynuclear metal carbonyl complexes, which are used as the precursors for the generation of the Sn-containing nanocluster catalysts anchored onto silica[28] (see Scheme 8.2). Incorporation of Sn into PGM catalysts is an effective, cheaper and convenient way of utilizing new catalysts.

Examples of the efficiency of incorporating Sn into nanocluster bimetallic and trimetallic clusters are given in Figures 8.12, 8.13 and 8.14, all taken from the work of the author and Adams and his colleagues.[26–28,34]

8.6 Quantum Computational Insights

Because it is extremely difficult to determine the structure (let alone the actual dynamics during catalytic turnover) of the nanocluster

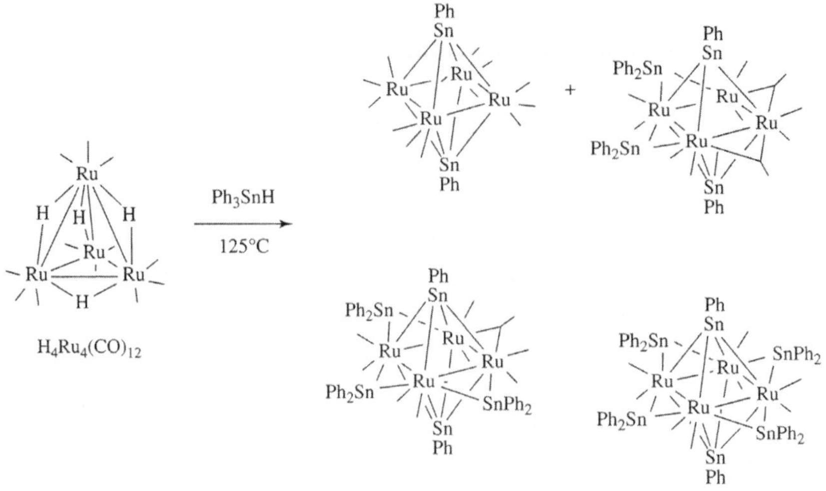

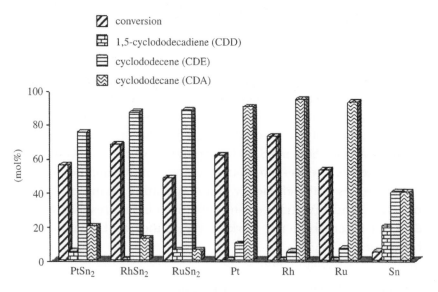

Scheme 8.2 The conversion of a Ru$_4$ carbonylate (left) to analogues that are rich in Sn adducts, these being the stepping stones for the production of novel bimetallic Ru-Sn clusters. (After Adams and Trufan.[34])

Figure 8.12 The effect of tin on the selective hydrogenation of cyclododecatriene (CDT) using anchored monometallic and bimetallic MSn$_2$ nanocluster catalysts. (After Adams and Thomas, adapted from Thomas *et al.*[28])

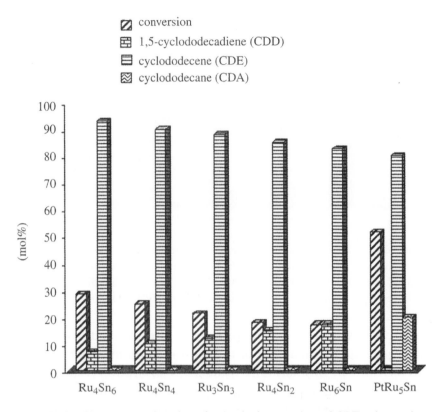

Figure 8.13 Chart comparing the selective hydrogenation of CDT using various bimetallic Ru-Sn and Ru_5PtSn nanocluster catalysts. (Adapted from Thomas *et al.*[28])

bi- and trimetallic catalysts under operational, and even under *ex situ* conditions, quantum computational approaches are a major aid in increasing our understanding of the nature of these high-performance catalysts. Whilst it is feasible to record *in situ* (using X-ray absorption spectroscopy) the immediate atomic environment (i.e. bond distances and coordination numbers) of each of the metals that constitute a bimetallic cluster catalyst — as was done with the $Ru_{12}Cu_4C_2$ and $Ru_{10}Pt_2C_2$ entities (see Figures 8.9 and 8.11) — by recording the core-electron and extended X-ray fine structure,[35,36] it is, as yet, very difficult to track *at the requisite degree of time resolution* the dynamic (or fluxional) aspects of these entities. However,

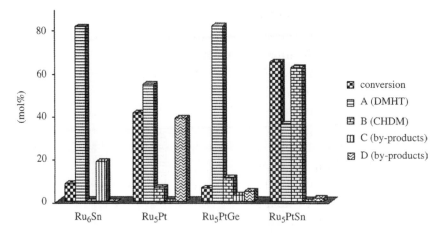

Figure 8.14 Chart comparing the activity and selectivity of the Ru_5PtSn catalyst (far right) with those of Ru_5PtGe, Ru_5Pt and Ru_6Sn nanocatalysts for the hydrogenation of dimethyl terephthalate (DMT) to cyclohexanedimethanol (CHDM). (Adapted from Hungria *et al.*, *Angew. Chem. Int. Ed.*, **45**, 4782 (2006).)

progress has been made in recovering the detailed static (time-averaged) atomic structure of the individual entities by combining (as was done by Bromley *et al.*[37]) the X-ray absorption results with an amalgam of molecular mechanics, molecular dynamics and density functional theory (DFT) calculations. The structures shown in Figures 8.1 and 8.9 were derived in this manner. (It is to be noted that the interstitial carbon atoms can also be located by utilizing the procedures used by Bromley *et al.*)

An admirable attempt has also been made[38] to probe the structure of Ru_3Sn_3 nanoclusters using scanning tunnelling microscopy. Triangular features were seen in the images; but the resolution was not adequate to identify the geometrical arrangement of the constituent elements.

Aberration-corrected high-resolution electron microscopy (AC-HREM)[39–41] still holds promise that it may be able to extract the desired structural and electronic characteristics of bimetallic and trimetallic nanoclusters, for it is already known that, under ideal imaging conditions,[17,40] it is possible to identify single atoms of PGMs attached on a light (low atomic number) support.

But beam-damage problems are often acute, and the migration of nanoclusters across the surface of the support often intensifies the difficulty of atomic imaging. We return again later to aspects of imaging nanoclusters by HREM. It is conceivable, but not yet proven, that the sub-threshold dosage method, recently described by Zewail *et al.*,[42] may prove helpful here.

Given the magnitude of these experimental difficulties associated with "interrogating" the nanoclusters, coupled to the fact that the clusters themselves are not distributed over the nanoporous SiO_2 support in a crystallographically ordered manner, one must resort to techniques such as DFT, which is an approach initiated by Grönbeck and the author.[43] We next outline the results so far retrieved from such an approach.

8.6.1 *The computational method*

DFT was applied with the gradient-corrected exchange-correlation functional according to Perdew *et al.*[44] The one-electron Kohn–Sham orbitals were expanded in a localized numerical basis set.[45] A pseudopotential was applied for Ru, Pt and Sn to describe the interaction between the valence electrons and the core, as described in Grönbeck and Thomas,[43] the following valence electrons being included, respectively, for Ru, Pt and Sn: $4s^2\, 4p^6\, 4d^7\, 5s^1$, $5s^2\, 5p^6\, 5d^9\, 6s^1$ and $4d^{10}\, 5s^2\, 5p^2$. The Kohn–Sham equations were solved self-consistently using an integration technique based on weighted overlapping spheres centred at each atom.[43]

8.6.2 *Assessing the structure and electronic properties of Ru$_5$PtSn in the gas phase and when supported on silica (cristobalite)*

Searches for stable isomers were performed by structural relaxation of a large number of initial configurations. About 20 structures (in different spin configurations) were relaxed for Ru$_5$PtSn as well as for Ru$_5$PtCSn. The structures were considered optimized when the largest gradient was smaller than 0.002 Ha Å^{-1} and the change in energy per optimization step was smaller than 10^{-5} Ha. The performance and reliability of this approach was checked by calculation for the Ru_2 dimer. The spectroscopic constants, bond length, binding energy and vibrational frequency were calculated to be 2.29 Å, 2.38 eV and 330 cm^{-1}, respectively, which are in good agreement with the available experimental data of the binding energy and the vibrational frequency, namely 2.0 ± 0.2 eV and 347 cm^{-1}, respectively.

8.6.2.1 *Gas-phase clusters*

The geometry-optimized structures for the gas-phase systems are reported in Figure 8.15. For the precursor carbonylate $Ru_5Pt(CO)_{15}(\mu\text{-}SnPh_2)(\mu\text{-}C)$, only one isomer was considered. For Ru_5PtCSn and Ru_5PtSn, the lowest two energy isomers were considered.

In the $Ru_5Pt(CO)_{15}(\mu\text{-}SnPh_2)(\mu\text{-}C)$ structure, as determined by DFT, the Ru-Ru distances are, on average, 2.95 Å, which is slightly longer than the experimental value of 2.90 Å. Also the average Ru-C distance computed is slightly longer than the experimental values: 2.09 Å compared with 2.06 Å. The cluster undergoes pronounced relaxation when the carbonyl and other ligands are removed. With complete removal, the carbido ligand is not stable in the interstitial position. Other computed incidental features of the "naked" gas-phase Ru_5PtCSn and $Ru_{15}PtSn$ clusters are of no direct relevance here.

The electronic DOS for the ground-state structures are shown in Figure 8.16. Projection of the Kohn–Sham orbitals onto the atomic states of the metal atoms (Ru, Pt and Sn) are shown in the red shaded areas. We note that:

- The carbonylated precursor cluster has a large HOMO-LUMO separation of *ca* 2 eV (see Figure 8.16(a)). This is to be expected for a stable molecular system. Energy states close to the HOMO-LUMO states are mainly of metal character.
- The Ru_5PtCSn clusters (Figure 8.16(b)) has a small HOMO-LUMO separation of 0.3 eV, which signifies a flexibility with respect to preferred spin multiplicity, and should clearly facilitate hydrogenation and dehydrogenation (as is

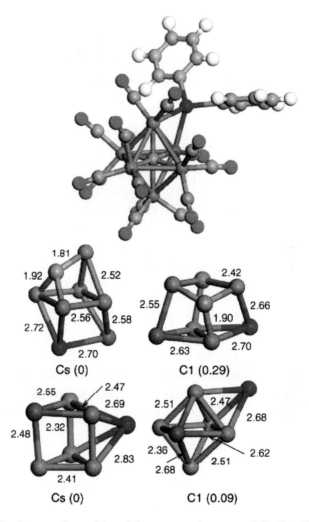

Figure 8.15 Structural models of low-energy isomers of Ru₅Pt (CO)₁₅(μ-C); Ru5PtCSn; and RuPt₅Sn. For the naked clusters, two low-energy isomers are reported. The point group symmetry is shown together with the relative energies in eV. Selected inter-atomic distances are given in Å. Colour code: Ru, dark green; Pt, blue; Sn, purple; C, grey; O, red; and H, white. (After Grönbeck and Thomas.[29])

observed experimentally). The ground state is calculated to be a triplet state, but it is essentially degenerate with a singlet configuration.

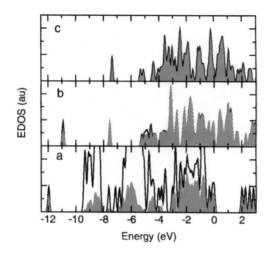

Figure 8.16 Electronic density of states for (a) $Ru_5Pt (CO)_{15}(\mu\text{-}SnPh_2)(\mu\text{-}C)$; (b) Ru_5PtCSn; and (c) Ru_5PtSn. The DOS is obtained by a 0.15 eV Gaussian broadening of the one-electron Kohn–Sham energies. The red shaded regions correspond to projection on metal (Pt, Cu and Sn) states. (DOS reported with respect to the HOMO level.)[29]

- The DOS for Ru_5PtSn (Figure 8.16(c)) has clear similarities with that of Ru_5PtCSn. The HOMO-LUMO gap is again small (0.2 eV) and the ground state is a triplet.

8.6.2.2 *A Ru_5PtCSn cluster supported on cristobalite (SiO_2)*[43]

Figure 8.17 shows the stable configuration of Ru_5PtCSn on the (001) face of α-cristobalite, which was initially covered with silanol groups, SiOH, just as they are in mesoporous silica. It transpires (computationally) that the interaction between the cluster and the oxygen atoms is so strong that H_2 is desorbed in the process. To be specific, the reaction

$$Ru_5PtCSn + (OH)_2 = Si< \;\rightarrow\; Ru_5PtCSn <^{O-S_1}_{O-S_1} + H_2$$

was computed to be exothermic by 2.7 eV; and the binding energy of the cluster to the silica surface is very high, 7.0 eV.[43] The cluster

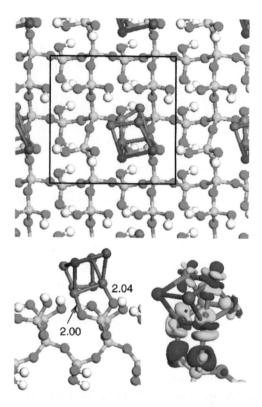

Figure 8.17 Upper part: top view of Ru$_5$PtCSn cluster supported on SiO$_2$. The surface cell is indicated. Lower part: side view and charge density difference map. The isosurface is shown at ±0.05 e/Å^3. Blue (yellow) isosurface denotes charge gain (depletion). Selected inter-atomic distances are given in Å. (After Grönbeck and Thomas.[29])

is linked to the surface via Ru-O bonds, with bond distances of 2.00 Å and 2.04 Å; and the configuration shown in Figure 8.17 is clearly preferred over all other (eight) variants that were computed. In view of the well-known oxophilicity of Sn, it is rather surprising that the nanocluster is linked to the silica support by two Ru-O-Si bonds. The computed Ru-O bond length (2.0 Å) is in reasonable agreement with the only known, experimentally determined[32] Ru-O bond length, namely 1.95 Å.

The character of the bond has a significant ionic component, with an accumulation of charge on the oxygen atoms. A Mulliken

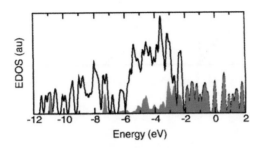

Figure 8.18   Electronic density of states (DOS) for Ru_5PtCSn supported on SiO_2. The DOS is obtained as in Figure 8.16. Red shaded region corresponds to projection onto Ru, Pt, Sn and C atoms. (After Grönbeck and Thomas.[29])

charge analysis indicated a charge transfer of about 0.5 electrons; and, interestingly, the main charge redistribution within the cluster is restricted to the Ru atoms forming the central unit.

The electronic DOS for the supported cluster is shown in Figure 8.18. Here, the area in red corresponds to projection onto the nanoparticle (including the C atom). The states close to the Fermi level are connected with the cluster; and states down to 2 eV below the Fermi level have a predominantly cluster character. The states with weights on C and Sn show small shifts with respect to the gas-phase analogue (*cf* Figure 8.16). Whereas the gas-phase cluster is a triplet, the singlet state is preferred over the triplet for the supported cluster by 0.1 eV.

8.6.3   *Quantum insights into the structure and densities of states of Ru_nSn_n (n = 3 to 6) clusters in the gas phase*

Using the DFT methods outlined earlier, Grönbeck *et al.*[46] have found that, in so far as structural considerations go, Ru atoms invariably constitute the core regions in Ru_3Sn_3, Ru_4Sn_4, Ru_5Sn_5 and Ru_6Sn_6 clusters in the energy-minimized structures. This is shown in Figure 8.19.

So far as the densities of states are concerned, there are only relatively minor differences between the distribution of energies as between the $(RuSn)_n$ when $n = 3$, 4, 5 or 6. Likewise the position of

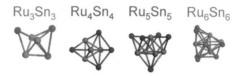

Figure 8.19 Computed, energy-minimized structures (in the gas phase) of Ru_3Sn_3, Ru_4Sn_4, Ru_5Sn_5 and Ru_6Sn_6. (After Grönbeck *et al.*[46])

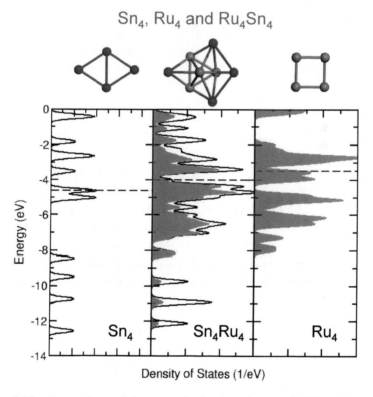

Figure 8.20 Comparison of (computed) density of states (DOS) of electronic energy in the energy-stabilized clusters of Sn_6, Ru_6 and Ru_6Sn_6. (After Grönbeck *et al.*[46])

the HOMO levels is similar in all these clusters (see Figure 8.20). When the DOS plots of the Ru_4Sn_4 clusters are compared with those of bulk Ru and bulk Sn (Figure 8.21), it is clear that the discrete energy levels of the clusters stand in sharp contrast to the continuous bands of energies of the two bulk metals, a conclusion

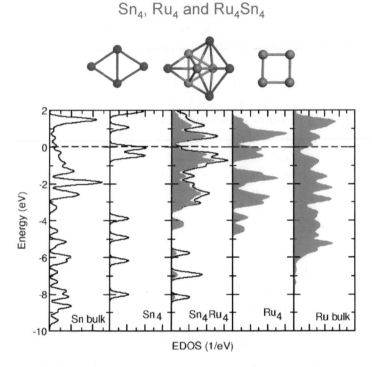

Figure 8.21 Comparison of differences in the continuous nature of the DOS between both bulk Sn and bulk Ru and of the discrete four- and eight-atom corresponding clusters. (After Grönbeck *et al.*[46])

unsurprising from earlier illustrations (see Figure 8.1) and discussions given in this chapter.

Work currently in progress aims to explore how these well-defined, energy-minimized bimetallic clusters behave with respect to the adsorption of H_2 — and, ultimately, to reaction pathways involving a simple catalytic reaction.

8.7 Comparisons with Nanocluster Catalysts Involving Gold, Platinum, Palladium and Iridium

The exceptionally high catalytic activity of finely divided metals, especially Au, Pt and Pd, has elicited enormous attention ever since

Haruta *et al.*[47] reported the totally unexpected catalytic performance of nanoparticle Au for the oxidation of CO to CO_2. Concerns about auto-exhaust pollution, for example, prompted the speculation that, provided they possessed intrinsic longevity, supported Au nanoparticles, since they catalyse the oxidation of CO in air at sub-zero temperatures (centigrade), could figure as components of future auto-exhaust and other environmentally important systems.

As emphasized in Section 8.1, we are not concerned here with the well-established and extensively studied remarkable catalytic performance of *nanoparticle* gold (first discovered by Haruta *et al.* and subsequently confirmed by many others.[48–52] Instead we focus on those reports that deal with nanocluster Au, Pt and Pd in which, increasingly, there is strong evidence of exceptional activity associated with minute clusters (perhaps single atoms or ions) of these metals. A summarizing account has been given[51] elsewhere of the nature and possible causes of the onset of high catalytic activity of Au nanoparticles in CO oxidation as a function of diameter of the nanoparticles[50] (see Figure 8.22).

Several workers, notably Flytzani-Stephanopoulos,[53] Mavrikakis,[54] Tsukuda[55] and colleagues, as well as this author, Grönbeck and Gai,[56]

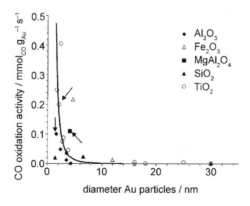

Figure 8.22 Plot showing the onset of high catalytic activity in CO oxidation in nanoparticles of gold.[50] (The three points marked by arrows are from the measurements of Lopez *et al.*,[50] the remainder are from the work of several other investigators.)

have reported convincing evidence that small sub-nanometer clusters are exceptionally active catalysts on a variety of supports including CeO_2, SiO_2, Al_2O_3, TiO_2 and hydroxyapatite. For example, Flytzani-Stephanopoulos *et al.*[54] came to the conclusion that, for the class of nanostructured Au/CeO_2 or Pt/CeO_2 catalysts, metal nanoparticles do not participate in the water-gas shift (WGS) reaction: $CO + H_2O \rightleftarrows CO_2 + H_2$.

The work of Flytzani-Stephanopoulos, Mavrikakis *et al.* raises the prospect that their atomically dispersed Pt catalysts might, in future, form the basis of a viable (on-board) H_2 source for vehicular fuel cells.[57] The authors maintained that Pt nanoparticles, non-metallic Au or Pt, strongly associated with the oxygen atoms of the CeO_2 support (through ionic-covalent Au-O-Ce bonds), are responsible for the activity. Corroborative evidence for this view has recently been presented by Park *et al.*[58]

In the paper by Zhai *et al.*[54] it has been shown that alkali-metal ions (Na^+ or K^+), added in small amounts, activate atomically dispersed Pt on Al_2O_3 or SiO_2 for the WGS reaction. They provide strong evidence that metallic Pt is not involved in the catalysis because the so-called white-line (X-ray absorption) intensity, in their *in situ* X-ray near-edge absorption spectra, clearly signifies an electron-deficient metal in the supported catalyst which they designate as Pt-$3K$-SiO_2 (and also Pt-$3Na$-SiO_2). These workers provide evidence that alkali-metal ions associated with surface OH groups are activated by CO at a temperature as low as 100 °C in the presence of atomically dispersed Pt. The nature of the active centre (which is single-site but multinuclear as with the bimetallic nanocluster catalysts described earlier in this chapter) is thought to be that represented in Figure 8.23. High-resolution electron micrographs taken by the Flytzani-Stephanopoulos group (see Figure 8.24) certainly are in line with her and her colleagues' interpretations. Thanks to aberration-corrected high-resolution electron micrographs, it is now possible to see single atoms[39,56,59] of Au and nanoclusters of less than 10 Å diameter (see Figure 8.25).

A surprising recent result concerning the nature of the widely used hydrogenation catalyst composed of Pt supported on γ-Al_2O_3 is

Atomically-Dispersed, Alkali-Metal-Ion-Stabilized
Platinum Catalysts are Exceptionally Active in the
Water-Gas Shift Reaction for Generating Hydrogen.
(M. Flytzani-Stephanopoulos, M. Mavrikakis *et al*, *Science*, **2010**, *329*, 1633)

Figure 8.23 Schematic depiction of how the Alkali-Metal-Ion-Stabilized Pt-OH$_x$ species (centre) catalyses the water-gas shift reaction at low temperature. (Provided by Flytzani-Stephanopoulos *et al.*[54])

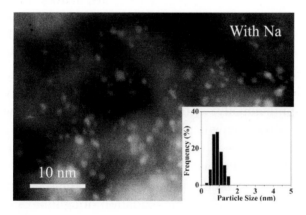

Figure 8.24 Typical HREM image of the Na$^+$-ion-stabilized Pt-OH$_x$ nanocluster catalysts designed by Zhai *et al.*[54] (Provided by Flytzani-Stephanopoulos *et al.*[54])

contained in the work of Kwak *et al.*[60] Using a combination of AC-HREM and ultra-high magnetic field MHS-NMR (of the ^{27}Al nucleus), these workers showed that coordinatively unsaturated pentacoordinate (designated Al$_5^{3+}$) centres present on the {100} faces of the Al$_2$O$_3$ surface are the sites at which Pt atoms are

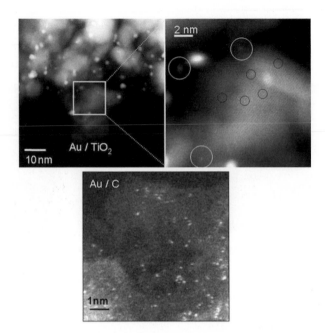

Figure 8.25 Aberration-corrected (AC) HREM images of nanoclusters of Au on (top) a titania and (bottom) an activated carbon support. Sub-nanometer clusters (white circles), which are approximately 5 Å in diameter and contain roughly ten Au atoms, and individual Au atoms (black circles) are observed.[56]

anchored (individually) in the $Pt/\gamma\text{-}Al_2O_3$ catalyst. At low coverages (of Pt), the active catalytic phase is atomically dispersed on the support surface, such that $(Pt/Al_5^{3+} \sim 1)$. At higher coverages a two-dimensional raft of Pt forms in the PtO phase.

The work of Tsukuda and colleagues[55] on size-controlled Au clusters in the selective aerobic oxidation of cyclohexane at 150 °C is particularly relevant here. They recently reported on the catalytic performance — with turnover frequencies (TOFs) in the range of 10^4 h^{-1} — of Au nanoclusters (freed from their passivating tripeptide molecules) supported on hydroxyapatite $(Ca_{10}(PO_4)_6(OH)_2)$ (HAP). They conducted their oxidation in the presence of traces of tertiary butyl hydroperoxide (TBHP), so as to generate the free radicals required for this oxidation (compare with Section 5.3.2). Their electron micrographs (see Figure 8.26) show how sharply

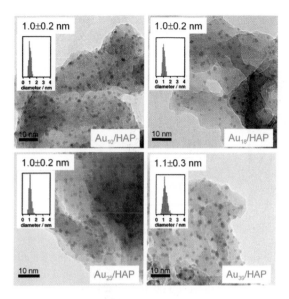

Figure 8.26 Typical HREM images of sub-nanometer Au clusters supported on hydroxyapatite. The Au_{10}, Au_{18}, Au_{25} and Au_{39} clusters were obtained by size-controlled methods. (After Tsukuda *et al.*[55])

defined the size distributions of the Au nanocatalysts are; and Figure 8.27 reveals that the TOFs pass through a maximum at 39 atoms of Au. This is an intriguing result which awaits full theoretical interpretation.

8.7.1 *Nanocluster catalysts of palladium and iridium*

Organic chemists have long suspected that in Suzuki, Negishi and Heck Pd-catalysed cross couplings in organic syntheses, the best Pd catalyst seems to be one which possesses the smallest possible size. In Reetz's colourful words, "homeopathic quantities of Pd will suffice".[61] In their case study involving Pd-containing perovskites (of formula $LaFe_{0.67}Co_{0.38}Pd_{0.05}O_3$), Ley and co-workers[62] concluded that Pd species of atomic dimension were crucial in governing the activity of this nominally heterogeneous catalyst, if only because it might serve as a "precatalyst" for what, in effect, turns out to be a solution-

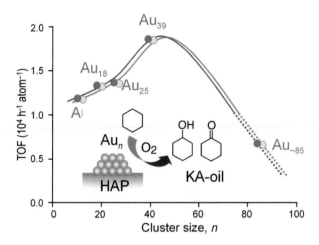

Figure 8.27 The catalytic activity of Au nanoclusters in the selective aerobic oxidation of cyclohexane to a mixture of cyclohexanol and cyclohexanone passes through a maximum when there are 39 atoms in the cluster. (After Tsukuda *et al.*[55])

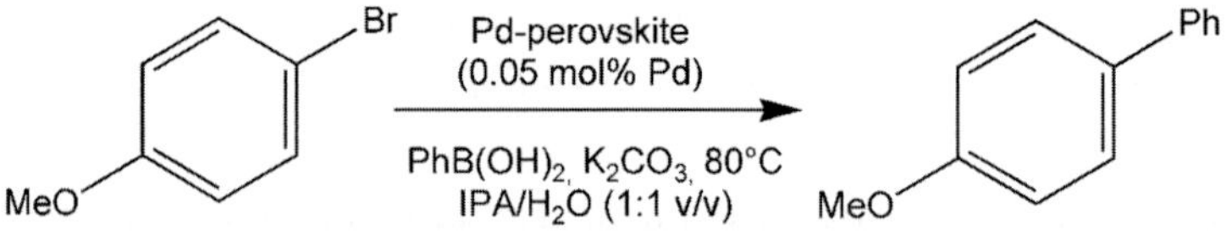

Figure 8.28 A typical Suzuki (homogeneous) coupling reaction involving phenylboronic acid and 4-bromoanisole. (After Smith *et al.*[63])

phase mechanism for the well-known Suzuki reaction of 4-bromoanisole with phenylboronic acid[63] (see Figure 8.28).

Very few detailed *in situ* studies of catalyst performance have been carried out on a single (or a few) well-defined metal nanoclusters. However, Gates and co-workers,[64] as described earlier (see Section 4.10), have made significant progress in identifying the nature of the single-site multinuclear catalysts Ir_4 and Ir_6 supported on different oxides, namely γ-Al_2O_3 and MgO. They came to some firm conclusions:

- Metal atoms at the nanocluster-support interface may be positively charged and the interactions with the support may stabilize the dispersion of the metal.

- Supports act as ligands, interacting with the metal through metal-oxygen bonds.
- Cluster size effects are significant, even for so-called structure-insensitive reactions. (This harmonizes with the recent results of Tsukuda *et al.*[55])
- Comparison of calculated structures with the experimentally determined (by EXAFS) structures (of nanocluster hydrogenation catalysts) indicates the possibility that high concentrations of hydrogen may exist within the supported nanoclusters.

What is abundantly clear is that joint studies involving, on the one hand, *in situ* structural monitoring of nanocluster structure during catalytic turnover and, on the other, more sophisticated quantum chemical techniques, are called for.

8.7.2 *The role of the catalyst support*

This is a subject much studied, for it is well known that readily reducible oxide supports (e.g. TiO_2 and CeO_2, each of which may take up gross non-stoichiometry upon exposure to H_2 at modest temperature) behave quite differently, with the same supported metal, from those oxides that are reducible with difficulty such as γ-Al_2O_3 and SiO_2. The overwhelming fraction of multinuclear single-site catalysts discussed in this chapter are supported on SiO_2. Recent work involving Au and Pt often use CeO_2 supports, so it is instructive to enquire more about the atomic structure of these supports. This can be done through the agency of AC-HREM; but it must be borne in mind that some of the features observed by this powerful microscopic technique (which entails rather fierce electron irradiation of the sample under investigation) may be the result of beam damage.

Shown in Figure 8.29 is a "dynamic" series of images (taken by my colleague Z. Saghi[65]) of a CeO_2 support, where it is seen that the tip of the specimen (consisting of a (100) face) is rather unstable. There is, in a sense, a fluidity about the surface of this support, which may also be present when the CeO_2 functions as a support for nanocluster and nanoparticle metal catalysts in auto-exhaust systems. (CeO_2 is one of the principal components of

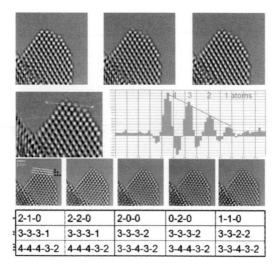

2-1-0	2-2-0	2-0-0	0-2-0	1-1-0
3-3-3-1	3-3-3-1	3-3-3-2	3-3-3-2	3-3-2-2
4-4-4-3-2	4-4-4-3-2	3-3-4-3-2	3-4-4-3-2	3-3-4-3-2

Figure 8.29 Dynamic AC-HREM images of a CeO_2 support showing a rather unstable (100) face (see top four micrographs, taken at 2 s intervals). A line profile through the second top layer of the image (left of middle set) shows successively one-, two-, three- and four-atom stacks. Using a single-atom column for calibration, we see that the line profiles at *, ** and *** of the image at the bottom left reflect the dynamism among surface atoms in the surface of CeO_2. (After Saghi.[65])

auto-exhaust catalysts.) It may well prove possible, using the environmental transmission electron microscope (E-TEM) devised by Gai, and now being modified to function in aberration-corrected mode, to probe the nanocluster-CeO_2 interface under near *in situ* conditions.

One of the inherent problems associated with HREM is the local heating that it inevitably injects into the specimen. This often leads to "crystallization" of small nanoclusters into larger, rather flat, objects, as we see in Figure 8.30 for a nanocluster Co supported on silica.

8.8 Envoi

For quantitatively understandable electronic reasons the performance of minute nanocluster catalysts far exceeds that of their

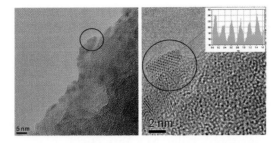

Figure 8.30 Small nanoclusters of Co, supported on SiO_2, "recrystallize" (as discussed in the text and Ref. 59) under electron irradiation. (The line profile in the top right-hand corner yields the inter-atomic distance between the Co atoms in the crystallined raft. (After Saghi.[65])

nanoparticle and bulk analogues: their LUMO-HOMO levels, being higher, facilitate the transfer of electrons to and from reacting surface species. Moreover, bimetallic nanocluster catalysts exhibit much greater activity than nanoclusters of the individual component. Apart from the intrinsic academic interest associated with these multinuclear single-site heterogeneous catalysts, especially in solvent-free hydrogenation reactions, they point the way towards the design of environmentally important new catalysts required in the age of sustainability and clean technology.

Most of the nanocluster catalysts described in this chapter have less than 20 metal atoms, and they are more akin to a "molecular metal", the precise electronic nature of which awaits further elucidation.[66] They do not possess well-developed crystallographic features, and are, in effect, at a smaller physical scale than the important, but larger, nanoscale catalysts that are amenable to the design of complex morphologies and pore structures.[67]

References

1. S. Schauermann, J. Hoffmann, V. Johanek, J. Hartmann, J. Libuda and H.-J. Freund. Catalytic activity and poisoning of specific sites on supported metal nanoparticles, *Angew. Chem. Int. Ed.*, **41**, 2532 (2002).

2. N. Nilius, N. Ernst and H.-J. Freund. Photon emission spectroscopy of individual oxide-supported silver clusters in a scanning tunnelling microscope, *Phys. Rev. Lett.*, **84**, 3994 (2000).

3. L. Osterlund, A. W. Grant and B. Kasemo. Lithographic techniques in nanocatalysis, in *Nanoscience and Technology of Nanocatalysis*, ed. U. Heiz and U. Landman, Springer, Berlin (2007).

4. J. H. Sinfelt. *Bimetallic Catalysts: Discoveries, Concepts and Applications* (Exxon Monograph), Wiley, New York (1983).

5. J. M. Thomas, B. F. G. Johnson, R. Raja, G. Sankar and P. A. Midgley. High-performance nanocatalysts for single-step hydrogenations, *Acc. Chem. Res.*, **36**, 20 (2003).

6. S. Hermans, R. Raja, J. M. Thomas, B. F. G. Johnson, G. Sankar and D. Gleeson. Solvent-free, low temperature, selective hydrogenation of polyenes using a bimetallic nanoparticle Ru-Sn catalyst, *Angew. Chem. Int. Ed.*, **40**, 1211 (2001).

7. R. Raja, T. Khimyak, J. M. Thomas, S. Hermans and B. F. G. Johnson. Single-step, highly active and highly selective nanoparticle catalysts for the hydrogenation of key organic compounds, *Angew. Chem. Int. Ed.*, **40**, 4638 (2001).

8. J. M. Thomas and G. N. Greaves. Probing solid catalysts under operating conditions, *Science*, **265**, 1675 (1995).

9. J. S. McIndoe and P. G. Dyson. *Transition Metal Carbonyl Cluster Chemistry*, Gordon and Breach Publishing, Amsterdam (2000).

10. D. Ozkaya, W. Zhou, J. M. Thomas, P. A. Midgley, V. J. Keast and S. Hermans. High-resolution imaging of nanoparticle bimetallic catalysts supported on mesoporous silica, *Catal. Lett.*, **60**, 113 (1999).

11. P. A. Midgley, E. P. W. Ward, A. B. Hungria and J. M. Thomas. Nanotomography in the chemical, biological and materials sciences, *Chem. Soc. Rev.*, **36**, 1477 (2007).

12. A. B. Hungria, R. Raja, R. D. Adams, B. Captain, J. M. Thomas, P. A. Midgeley, V. Golvenko and B. F. G. Johnson. Bimetallic Ru-Sn nanoparticle catalysts for the solvent-free selective hydrogenation of 1,5,9-cyclododecatriene to cyclododecene, *Angew. Chem. Int. Ed.*, **46**, 8182 (2007).

13. R. J. Davis and M. Boudart. Structure of supported Pd-Au clusters determined by X-ray absorption spectroscopy, *J. Phys. Chem.*, **98**, 5471 (1994).

14. J. M. Thomas, R. Raja, B. F. G. Johnson, S. Hermans, M. D. Jones and T. Khimyak. Bimetallic catalysts and their relevance to the hydrogen economy, *Ind. Eng. Chem.*, **42**, 1563 (2003).

15. R. Raja, G. Sankar, S. Hermans, D. S. Shephard, S. T. Bromley, J. M. Thomas, T. Maschmeyer and B. F. G. Johnson. Preparation and characterization of highly active bimetallic (Pd-Ru) nanocatalysts, *Chem. Commun.*, 1571 (1999).

16. J. M. Thomas and G. Sankar. *In situ* combined X-ray absorption spectroscopic and X-ray diffractometric studies of solid catalysts, *Top. Catal.*, **8**, 1 (1999).

17. J. M. Thomas, O. Terasaki, P. L. Gai, W. Zhou and J. M. Gonzalez-Calbet. Structural elucidation of microporous and mesoporous catalysts and molecular sieves by high-resolution electron microscopy, *Acc. Chem. Res.*, **34**, 583 (2001).

18. J. M. Thomas, P. A. Midgley, T. J. V. Yates, J. S. Barnard, R. Raja, I. Arslan and M. Weyland. The chemical application of high-resolution electron tomography: bright field or dark field?, *Angew. Chem. Int. Ed.*, **43**, 6745 (2004).

19. M. Weyland, J. M. Thomas, P. A. Midgley and B. F. G. Johnson. Z-contrast tomography: a technique in three-dimensional nanostructural analysis based on Rutherford scattering, *Chem. Commun.*, 907 (2001).

20. R. M. Rioux, H. Song, J. D. Hoefelmeyer, P. Yang and G. A. Somorjai. High-surface-area catalyst design: synthesis, characterization, and reaction studies of platinum nanoparticles in mesoporous SBA-15 silica, *J. Phys. Chem. B*, **109**, 2192 (2005).

21. J. W. Shabaker, D. A. Simonetti, R. D. Cortright and J. A. Dumesic. Sn-modified Ni catalysts for aqueous-phase reforming, *J. Catal.*, **231**, 67 (2005).

22. B. C. Gates. *Catalytic Chemistry*, Wiley, New York (1992).

23. J. M. Thomas and W. J. Thomas. *Heterogeneous Catalysis: Principles and Practice*, Wiley-VCH, Weinheim (1997).

24. M. Guidotti, V. Dal Santo, A. Gallo, E. Gianotti, G. Peli, R. Psaro and L. Sordelli. Catalytic dehydrogenation of propane over cluster-derived Ir-Sn/SiO$_2$ catalysts, *Catal. Lett.*, **112**, 89 (2006).

25. A. Siani, B. Captain, O. S. Alexeev, E. Stafyla, A. B. Hungria, P. A. Midgley, J. M. Thomas, R. D. Adams and M. D. Amiridis. Improved

CO-oxidation activity in the presence and absence of H_2 over cluster-derived $PtFe/SiO_2$ catalysts, *Langmuir*, **22**, 5160 (2006).

26. R. D. Adams, E. M. Boswell, B. Captain, A. B. Hungria, P. A. Midgley, R. Raja and J. M. Thomas. Bimetallic Ru-Sn nanoparticle catalysts for the solvent-free selective hydrogenation of 1,5,9-cyclodecatriene to cyclododecene, *Angew. Chem. Int. Ed.*, **46**, 8182 (2007).

27. R. D. Adams, D. A. Blom, B. Captain, R. Raja, J. M. Thomas and E. Trufan. Toward less dependence on platinum group metals: the merits of utilizing tin, *Langmuir*, **24**, 9223 (2008).

28. J. M. Thomas, R. D. Adams, E. M. Boswell, B. Captain, H. Grönbeck and R. Raja. Synthesis, characterization, electronic structure and catalytic performance of bimetallic and trimetallic nanoparticles containing tin, *Faraday Disc.*, **138**, 301 (2008).

29. H. Grönbeck and J. M. Thomas. Structural and electronic properties of a trimetallic nanoparticle catalyst: Ru_5PtSn, *Chem. Phys. Lett.*, **443**, 337 (2007).

30. P. Appleton and M. A. Wood (Eastman Chemicals). U.S. Patent No. 35414159 (1993).

31. J. M. Thomas, R. Raja, P. L. Gai, H. Grönbeck and J. C. Hernandez-Garrido. Exceptionally active single-site nanocluster multifunctional catalysts for cascade reactions, *Chem. Cat. Chem.*, **2**, 402 (2010).

32. J. M. Thomas, R. Raja, B. F. G. Johnson, T. J. O'Connell, G. Sankar and T. Khimyak. Bimetallic nanocatalysts for the conversion of muconic acid to adipic acid, *Chem. Commun.*, 1126 (2003). See also Editor's choice: Highlights of the Recent Literature, *Science*, **300**, 867 (2003).

33. R. Burch and L. C. Garla. Platinum-tin reforming catalysts: II. Activity and selectivity in hydrocarbon reactions, *J. Catal.*, **71**, 360 (1981).

34. R. D. Adams and E. Trufan. Ruthenium-tin cluster complexes and their applications as bimetallic nanoscale heterogeneous hydrogenation catalysts, *Phil. Trans. R. Soc. A*, **368**, 1473 (2010).

35. R. D. Cortright and J. A. Dumesic. Microcalorimetric, spectroscopic and kinetic studies of silica-supported Pt and Pt-Sn catalysts for isobutene dehydrogenation, *J. Catal.*, **148**, 771 (1994).

36. D. S. Shephard, T. Maschmeyer, G. Sankar, J. M. Thomas, D. Ozkaya, B. F. G. Johnson, R. Raja, R. D. Oldroyd and R. G. Bell. Preparation, characterisation and performance of encapsulated copper-ruthenium

bimetallic catalysts derived from molecular cluster carbonyl precursors, *Chem. Eur. J.*, **4**, 1214 (1998).

37. S. T. Bromley, G. Sankar, C. R. A. Catlow, T. Maschmeyer, B. F. G. Johnson and J. M. Thomas. New insights into the structure of supported bimetallic cluster catalysts prepared from carbonylated precursors: a combined density functional theory and EXAFS study, *Chem. Phys. Lett.*, **340**, 524 (2001).

38. F. Yang, E. Trufan, R. D. Adams and D. W. Goodman. Structure of molecular-sized Ru_3Sn_3 clusters on a SiO_2 film on Mo(112), *J. Phys. Chem. C*, **112**, 14233 (2008).

39. O. L. Krivanek, M. F. Chisholm, V. Nicolosi, T. J. Pennycook, G. J. Corbin, N. Dellby, M. F. Murfitt, C. S. Own, Z. S. Szilagyi, M. P. Oxley, S. T. Pantelides and S. J. Pennycook. Atom-by-atom structural and chemical analysis by annular dark-field electron microscopy, *Nature*, **464**, 571 (2010).

40. J. M. Thomas and P. A. Midgley. The merits of static and dynamic high-resolution electron microscopy for the study of solid catalysts, *Chem. Cat. Chem.*, **2**, 783 (2010).

41. J. M. Thomas and P. A. Midgley. The modern electron microscope: a cornucopia of chemico-physical insights, *Chemical Physics*, **385**, 1 (2011).

42. O.-H. Kwon, V. Ortalan and A. H. Zewail. Macromolecular structural dynamics visualized by pulsed dose control in 4D electron microscopy, *PNAS*, **108**, 6026 (2011).

43. H. Grönbeck and J. M. Thomas. Structural and electronic properties of a trimetallic nanoparticle catalyst: Ru_5PtSn, *Chem. Phys. Lett.*, **443**, 337 (2007).

44. J. P. Perdew, K. Burke and M. Ernzerhof. Generalized gradient approximation made simple, *Phys. Rev. Lett.*, **77**, 3865 (1996).

45. B. Delley. From molecules to solids with $DMol^3$ approach, *Phys. Rev. B*, **66**, 155125 (2002).

46. H. Grönbeck, J. M. Thomas *et al.* (work in progress).

47. M. Haruta, H. Sano, N. Yamada and T. Kobayashi. Novel gold catalysts for the oxidation of carbon monoxide at a temperature far below 0 °C, *Chem. Lett.*, **2**, 405 (1987).

48. L. Prati and M. Rossi. Gold on carbon as a new catalyst for selective liquid phase oxidation of diols, *J. Catal.*, **176**, 552 (1998).

49. C. Lemire, R. Meyer, S. Shaikhutdinov and H. J. Freund. Do quantum size effects control CO adsorption on gold nanoparticles?, *Angew. Chem. Int. Ed.*, **43**, 118 (2004).

50. N. Lopez, T. V. W. Janssens, B. S. Clausen, Y. Xu, M. Marrikakis, T. Bliggard and J. K. Norskov. On the origin of the catalytic activity of gold nanoparticles for low-temperature CO oxidation, *J. Catal.*, **223**, 232 (2004).

51. P. P. Edwards and J. M. Thomas. Gold in a metallic divided state — from Faraday to present-day nanoscience, *Angew. Chem. Int. Ed.*, **46**, 5480 (2007).

52. A. Corma, I. Dominguez, T. Rodenas and M. J. Sabater. Stabilization and recovery of gold catalysts in the cyclopropanation of alkenes within ionic liquids, *J. Catal.*, **259**, 26 (2008).

53. Q. Fu, H. Saltsburgh and M. Flytzani-Stephanopoulos. Active non-metallic Au and Pt species on ceria-based water-gas shift catalysts, *Science*, **301**, 935 (2003).

54. Y. Zhai, D. Pierre, R. Si, W. Deng, P. Ferrin, A. U. Nilekar, G. Peng, J. A. Herron, D. C. Bell, H. Saltsburg, M. Mavrikakis and M. Flytzani-Stephanopoulos. Alkali-stabilized Pt-OH$_x$ species catalyze low-temperature water-gas shift reactions, *Science*, **329**, 1633 (2010).

55. Y. Liu, H. Tsunoyama, T. Akita, S. Xie and T. Tsukuda. Aerobic oxidation of cyclohexane catalyzed by size-controlled Au clusters on hydroxyapatite: size effect in the sub-2 nm regime, *ACS Catal.*, **1** (2011), DOI: 10.1021/cs100043.

56. J. M. Thomas, R. Raja, P. L. Gai, H. Grönbeck and J. C. Hernandez-Garrido. Exceptionally active single-site nanocluster multifunctional catalysts for cascade reactions, *Chem. Cat. Chem.*, **2**, 402 (2010).

57. J. M. Thomas. An exceptionally active catalyst for generating hydrogen from water, *Angew. Chem. Int. Ed.*, **50**, 49 (2011).

58. J. B. Park, J. Graciani, J. Evans, D. Stacchiola, S. Ma, P. Liu, A. Nambu, J. Fernández Sanz, J. Hrbek and J. A. Rodriguez. High catalytic activity of Au/CeO$_x$/TiO$_2$(100) controlled by the nature of the mixed-metal oxide at the nanometer level, *PNAS*, **106**, 4975 (2009).

59. E. P. W. Ward, I. Arslan, P. A. Midgley, A. Bleloch and J. M. Thomas. Direct visualisation, by aberration-corrected electron microscopy, of the crystallisation of bimetallic nanoparticle catalysts, *Chem. Commun.*, 5805 (2005).

60. J. H. Kwak, J. Hu, D. Mei, C. W. Yi, D. H. Kim, C. H. F. Peden, L. F. Allard and J. Szanyi. Coordinatively unsaturated Al^{3+} centers as binding sites for active catalyst phases of Pt on γ-Al_2O_3, *Science*, **325**, 1670 (2009).

61. M. T. Reetz. Private communication (2009).

62. S. P. Andrews, A. F. Stepan, H. Tanaka, S. V. Ley and M. D. Smith. Heterogeneous or homogeneous? A case study involving palladium-containing perovskites in the Suzuki reaction, *Adv. Synth. Catal.*, **347**, 647 (2005).

63. M. D. Smith, A. F. Stepan, C. Ramarao, P. E. Brennan and S. V. Ley. Palladium-containing perovskites: recoverable and reusable catalysts for Suzuki couplings, *Chem. Commun.*, 2652 (2003).

64. A. Uzun, D. A. Dixon and B. C. Gates. Prototype supported metal cluster catalysts: Ir_4 and Ir_6, *Chem. Cat. Chem.*, **3**, 95 (2011).

65. Z. Saghi. Private communication. See also J. M. Thomas, Z. Saghi and P. L. Gai. Can a single atom serve as the active site in some heterogeneous catalysts? *Top. Catal.*, **54**, 588 (2011).

66. J. M. Thomas. Heterogeneous catalysis: enigmas, illusions, challenges, realities and emergent strategies of design, *J. Chem. Phys.*, **128**, 182502 (2008).

67. J. Kleis, J. K. Norskov *et al.* Finite size effects in chemical bonding: from small clusters to solids. *Catal. Lett.*, **141**, 1067 (2011).

68. W. Zhou, J. M. Thomas, B. F. G. Johnson, D. Ozkaya, T. Maschmeyer, R. G. Bell and Q. Ge. Illustration of how an organic cation (PPN^+) may be incorporated along with a carbonylate $(H_2Ru_{10}(Co)_{25})^{2-}$ inside a silica mesopore, *Science*, **280**, 705 (1998).

APPENDIX I

REFERENCE WORKS DEALING WITH GREEN CHEMISTRY, CLEAN TECHNOLOGY AND SUSTAINABILITY

Several recent texts have covered the use and importance of heterogeneous catalysts — but very few of the single-site variety — in the general thrust of this monograph. Texts that are recommended for further general reading are the following:

- F. Cavani, G. Centi, S. Perathoner and F. Trifiro (eds). *Sustainable Industrial Chemistry: Principles, Tools and Industrial Examples,* Wiley-VCH, Weinheim (2009).
- R. H. Crabtree and P. T. Anastas (eds). *Green Catalysis, Vol. 2: Heterogeneous Catalysis,* Wiley-VCH, Weinheim (2009).
- R. A. Sheldon, I. Arends and U. Hanefeld. *Green Chemistry and Catalysis,* Wiley-VCH, Weinheim (2007).
- R. A. van Santen and M. Neurock. *Molecular Heterogeneous Catalysis: A Conceptual and Computational Approach,* Wiley-VCH, Weinheim (2006).

Other texts that deal in part with single-site heterogeneous catalysts include:

- G. Ertl, H. Knözinger, F. Schüth and J. Weitkamp (eds). *Handbook of Heterogeneous Catalysis* (in eight volumes), Wiley-VCH, Weinheim (2008).

- P. Barbaro and C. Bianchini (eds). *Catalysis for Sustainable Energy Production,* Wiley-VCH, Weinheim (2009).
- J. Čejka, A. Corma and S. Zones (eds). *Zeolites and Catalysis: Synthesis, Reactions and Applications,* Wiley-VCH, Weinheim (2010).
- P. Barbaro and F. Liguori (eds). *Heterogenized Homogeneous Catalysts for Fine Chemicals Production,* Springer Science, Berlin.

Two texts worthy of consultation that are devoted to catalysis in oxidations are:

- J.-E. Bächvoll (ed.). *Modern Oxidation Methods,* Wiley-VCH, Weinheim (2004).
- N. Mizuno (ed.). *Modern Heterogeneous Oxidation Catalysis,* Wiley-VCH, Weinheim (2009).

In addition to the well-established journals that focus on catalysis in general — and which now give increasing prominence to single-site heterogeneous catalysts — such as *Catalysis Today, Catalysis Letters, Topics in Catalysis* and *Applied Catalysis,* two new journals have recently (2011) appeared from the American Chemical Society (ACS) and the Royal Society of Chemistry (RSC).

The RSC has been publishing *Green Chemistry* since 1998. Some special issues of *Topics in Catalysis* have been devoted to general and specific aspects of this monograph. See, for example:

- G. Giambastini and C. Bianchini (eds). Homogeneous, Single-Site Heterogeneous and Nanostructured Catalysts for Sustainable Development, *Top. Catal.,* **40**, 1–4 (2006).
- T. J. Marks and P. C. Stair (eds). The Interface Between Heterogeneous and Homogeneous Catalysis, *Top. Catal.,* **34**, 1–4 (2005).
- K. Smith and J. M. Thomas (eds). New Solid Catalysts for Clean Technology, *Top. Catal.,* **52**, 12 (2009).

A special issue of *Annual Reviews of Materials Science* (Vol. 35, 2005) dealt with materials design and chemistry of environmentally acceptable catalysts. Some of the key papers of that issue, relevant to single-site heterogeneous catalysts, were the following:

- G. J. Hutchings. Heterogeneous asymmetric catalysts: strategies for achieving high enantioselection, *Annu. Rev. Mater. Sci.*, **35**, 143–208 (2005).
- F. Schüth. Engineered porous catalytic materials, *Annu. Rev. Mater. Sci.*, **35**, 209–314 (2005).
- J. M. Thomas and R. Raja. Designing catalysts for clean technology, green chemistry and sustainable development, *Annu. Rev. Mater. Sci.*, **35**, 315–350 (2005).
- H. O. Pastore, S. Coluccia and L. Marchese. Porous aluminophosphates: from molecular sieves to designed acid catalysts, *Annu. Rev. Mater. Sci.*, **35**, 351–396 (2005).

The following articles are also relevant:

- P. N. R. Vennestrom, C. M. Osmundsen, C. H. Christensen and E. Taarning. Beyond petrochemicals. The renewable chemicals industry, *Angnew. Chem. Int. Ed.*, **50**, 10502 (2011).
- M. J. Climent, A. Corma and S. Iborra. Heterogeneous catalysts for the one-pot synthesis of chemicals and fine chemicals, *Chem. Rev.*, **111**, 1072–1133 (2011).
- L. Li, G. D. Li, C. Yan, X. Y. Mu, X. L. Pan, X. X. Zou, K. X. Wang and J. S. Chen. Efficient sunlight-driven dehydrogenative coupling of methane to ethane over Zn^+-modified zeolite, *Angew. Chem. Int. Ed.*, **50**, 8299–8303 (2011).
- M. Anpo and J. M. Thomas. Single-site photocatalytic solids for the decomposition of undesirable molecules, *Chem. Commun.*, 3273–3278 (2006).

INDEX